Lecture Notes
in Control and Information Sciences 319

Editors: M. Thoma · M. Morari

T0134763

Michael W. Hofbaur

Hybrid Estimation
of Complex Systems

With 69 Figures

Springer

Author

Michael W. Hofbaur, Ao. Univ.-Prof. Dr.
Graz University of Technology
Institute of Automation and Control
Inffeldgasse 16c/2
8010 Graz
Austria

ISSN 0170-8643

ISBN-10   3-540-25727-6 **Springer Berlin Heidelberg New York**
ISBN-13   978-3-540-25727-1 **Springer Berlin Heidelberg New York**

Library of Congress Control Number: 2005925478

Springer is a part of Springer Science+Business Media

springeronline.com

© Springer-Verlag Berlin Heidelberg 2005
Printed in Germany

Typesetting: Data conversion by author.
Final processing by PTP-Berlin Protago-TEX-Production GmbH, Germany
Cover-Design: design & production GmbH, Heidelberg
Printed on acid-free paper   89/3141/Yu - 5 4 3 2 1 0

To Xaver and Vitus

Xaver's perception of an advanced life support system in space

Preface

Modern technology is increasingly leading to complex artifacts with high demands on performance and availability. As a consequence, advanced automation and control methods play an important role in achieving these requirements. Key for automation and control is to know the operational state of an artifact in order to take the appropriate corrective measures. Often, however, it is not possible or desirable to measure all physical entities that determine the state of the system. So that the missing information has to be inferred from noisy measurements and a mathematical model of the physical system. The *Kalman filter* [61] is surely the most prominent example for this task of filtering and estimation. A Kalman filter utilizes a stochastic model of the system under investigation and estimates, in some optimal sense, the physical entities that determine the state of a system. When applied to real-world systems, however, one has to deal with changing operational conditions and a possible set of operational modes with significantly different dynamic behavior. A standard Kalman filter, however, can only cope with a single operational mode where its stochastic model matches the exhibited dynamic behavior of the system. A standard approach to deal with mode changes and abruptly varying dynamics is to utilize multiple filters, one for each operational mode, and apply some sort of hypothesis selection/merging to deduce the appropriate estimate of the system's state according to the individual estimates of the multiple filters. This *multiple model* approach executes filters for all possible modes concurrently so that any realistic real-time implementation limits the number of operational modes to a moderately large set of modes. This limitation, to capture a system by few operational modes only, is appropriate for many applications, however, todays complex artifacts, such as autonomous robots, space-probes, or production plants, impose new demands that go beyond the scope of standard multiple model estimation. Those systems are composed of many individual components and achieve their complex and high performance behavior by applying advanced control and automation schemes that utilize complex interactions among the components. This leads to overall systems with an overwhelming number of possible operational modes[1]. As a consequence, it becomes increasingly difficult to build estimation/monitoring systems that deal with the overwhelmingly large number of possible modes of modern complex artifacts.

Besides providing an estimate for the physical entities of a system, which are modeled as continuous variables, it is important to estimate the operational mode of the individual components as well. This is becoming increasingly important as more and more systems are built that operate autonomously, with little or almost no human interaction over long periods of time. So that some sort of automated self monitoring and diagnosis becomes vital for the reliable operation of these systems. This operation, however, ought not be done by breaking down the estimation problem into separate estimation tasks for the individual components of the system to be solved by individual estimators. Modern multivariate control schemes, such as nonlinear model predictive control, orchestrate the components of a system in a sophisticated interleaved way that utilizes complex interaction among the components of the system to achieve a desired operation. This system-wide interaction, however, would be missed by an estimation scheme that focuses on the system's components individually. Another consequence of this tight coupling among the system's components is that a fault in one component can manifest itself in another component. This makes it difficult, if not impossible, for a human operator to trace the operation in order to diagnose the system whenever a malfunction is experienced. Therefore, an automated monitoring and estimation capability, that can cope with the complexity of the system and capture the system-wide interactions, will be enormously helpful.

As already indicated above, the number of operational modes for a modern artifact can be overwhelmingly large. However, as we intend to track and *monitor* the operation of the system, it is worthwhile to consider additional *fault modes* in the course of estimation. A *failure mode effect analysis (FMEA)* is usually applied to identify possible faults/failures in the system and to anticipate their system-wide effects. Given today's complex systems, however, it might be reasonable to assume that it is difficult, if not impossible, to anticipate all possible fault modes and determine their system-wide effects. An exhaustive FMEA would reveal an unreasonably large number of fault modes, that would outnumber the already large number of operational modes, in particular, when taking multiple faults into account. As a consequence, even a highly focused estimation scheme would be overwhelmed by the large number of possible mode candidates that it has to consider in the course of estimation. This would impose an unrealistically large computational burden on a typical real-time computing environment. Furthermore, unanticipated faults do occur, in particular, whenever a system operates in a poorly specified environment (e.g. planetary rovers). Of course, it is good practice to define particular

---

[1] For instance, consider a moderately complex system that is composed of 10 components, where each component can exhibit one out of 5 operational modes. This leads to an overall system with $5^{10} \approx 10,000,000$ modes.

fault modes that might occur, so that an estimation/monitoring system can identify those pre-specified operational situations. However, it is also desirable to include the capability to cope with unanticipated operational situations. In this way, an estimation/monitoring system can maintain operational whenever parts of the system exhibit at an *unknown mode* of operation.

Today's complex artifacts are often built for long and durable operation. As a consequence, parts of it might get replaced or updated and it becomes increasingly difficult to maintain and update the estimation/monitoring system for the artifact. For instance, replacing a servo-valve in a production plant with one that exhibits a slightly different behavior should not trigger the necessity for an off-line system-wide redesign of a high fidelity estimator that takes system-wide interactions into account. However, it is reasonable to replace the model-fragment of the system's overall model that deals with the particular device and leave it to the estimation/monitoring system do deal with the changed situation.

Contributions

This outlines the scope of this work: we describe a novel model-based approach for on-line estimation that can deal with the complexity of modern automation systems, whilst being able to deal with (partially) unspecified operational situations. More specifically, we present:

Component-based modeling formalism: In order to deal with complex multi-component systems that exhibit hybrid discrete/continuous behaviors of their physical entities, we choose *hybrid automata* as general modeling paradigm. These automata describe physical systems in terms of a set of operational and fault modes, mode transitions, and a dynamic model for the physical entities for each specified mode. The novelty of our approach is that we combine a *probabilistic mode transition model* with *stochastic discrete-time difference equations* to capture real-world effects such as disturbances, sensor noise, and other non-deterministic effects. This leads to what we call a *probabilistic hybrid automaton (PHA)* that serves as the basic building block for our modeling paradigm. In terms of the overall model for a complex physical artifact, we take a *component-based* approach that models individual components of the physical artifact in terms of probabilistic hybrid automata and defines the interplay among them in terms of their concurrent composition. This leads to what we call a *concurrent probabilistic hybrid automaton*, or *cPHA* for short.

Focused hybrid estimation: Estimation for hybrid systems is generally difficult. This is due to the fact that an estimator has to consider all possible mode sequences with their associated continuous evolutions. This fact has particularly severe implications whenever the number of operational modes of the underlying system is very large so that standard methods from the field of multiple-model estimation cannot be applied anymore.

Our proposed hybrid estimation method deals with this class of complex systems. It carefully explores possible estimation hypotheses and focuses onto the likely estimates. For this purpose we carefully re-formulate the hybrid estimation problem as a best-first search problem and utilize advanced search techniques from the toolkit of Artificial Intelligence to solve this problem. This leads to an *any-time/any-space* hybrid estimation algorithm that is suitable for on-line execution within the context of an automation system.

Automated on-line filter design: Our hybrid estimation system utilizes a set of (extended) Kalman filters, one for each mode under consideration, to anticipate the continuous evolution of the system. Instead of using a pre-compiled set of (hand-crafted) filters, we provide a sophisticated model-based mechanism that deduces these filters on-line. This capability has two important implications. Firstly, there is no need to pre-compile a possibly prohibitively large number of filters. The limited computational resources of real-time systems, that execute the automation system, won't allow us to store and utilize them anyhow. Secondly, since the deduction process grounds upon the hybrid model for the physical artifact, we can easily update or modify the estimator to incorporate any modification of the physical artifact or components of thereof. We simply update the underlying component model. Hybrid estimation with the underlying automated filter design capability incorporates this change automatically.

Robustness: It is difficult, if not impossible, to anticipate all possible faults that can occur within a complex physical artifact. Un-anticipated situations do occur, and a wrong classification of thereof can cause severe implications such as the loss of control over a potentially dangerous automated system (chemical plant, power plant, airplane, automobile...). Our proposed hybrid estimation scheme takes this into account and provides a generic *unknown mode* that captures all un-anticipated modes of behavior. In this way, we obtain an estimation capability that can detect un-anticipated modes of operation, identify the impaired components or subsystem, and continue hybrid estimation in a degraded, but fail-safe manner.

Perspective: Autonomous Automation

This monograph deals almost exclusively with monitoring and estimation within the context of automation for complex systems. Its implication, however, should be seen within the wider perspective of advanced automation that deals with un-anticipated situations and an overwhelmingly large number of possible control strategies. Within the estimation task, we solve the problem in the sense that we apply advanced search techniques from the toolkit of modern AI to identify the most probable estimation hypothesis, given the observation and the underlying model. The dual problem would be to search for a suitable control strategy, given an abstract control goal and the hybrid model of the physical artifact under control. The algorithms that we developed to cope with the enormous amount of estimation hypotheses open a direction for future research for the dual control problem as well. Surely, they represent the starting point and will require additional enhancements with techniques from the toolkits of modern Control Theory, Automation, and Artificial Intelligence. However, on the long run, we expect to obtain an overall automation system that can reconfigure itself to adapt to changing environments and faults or even failures. The basis for such an advanced model-based automation paradigm, we could call it *autonomous automation*, is to robustly know the state of the physical artifact – the topic of this monograph.

Outline

The next Section (2) introduces our hybrid estimation paradigm on a concise basis that omits most of the low level detail. This should give a first glimpse of our proposed concept, motivate the overall design and the paradigms that we used within this work, and indicate problematic issues that require high-fidelity tools out of the toolbox of modern AI. It also indicates possible lines of future research and applications and reviews related and complimentary research. Section 3 and Sect. 4 provide a detailed treatment of our concept for hybrid estimation. Section 3 presents the basis for our hybrid estimation scheme, our hybrid modeling framework. Section 4 deals with the overall hybrid estimation task. It reviews the underlying theory and the existing approaches, and provides our key contributions – the focused hybrid estimation scheme, and the incorporation of the unknown mode (Sects. 4.5 and 4.6, respectively). Some examples are given in Sect. 5 to illustrate our proposed hybrid estimation scheme. The examples range from a relatively simple multi-component system, that allows us to compare our hybrid estimation scheme with traditional multiple model estimation algorithms, to the simulation study of an advanced life support system for a Martian space mission, as example for a process automation system.

Acknowledgments

This monograph is a revised version of my habilitation thesis [53] that I succesfully defended at the Faculty of Electrical Engineering at Graz University of Technology (TUG) in March 2004. It describes work that I pursued during the years 2000 - 2004 at the Institute of Automation and Control at TUG and at the Artificial Intelligence Laboratory at the Massachusetts Institute of Technology (MIT). Fruitful research is almost always done in collaboration with colleagues in academia and industry. The work presented in this book is no exception. In particular, I want to mention Brian C. Williams who supervised the work at MIT. Brian's guidance and support significantly shaped this research endeavor that builds upon his work on discrete model-based diagnosis. Many other people contributed in various important ways. I list them alphabetically and hope that they enjoy the friendship as much as I do:

Gautam Biswas, Seung Chung, Nico Dourdoumas, Stano Funiak, Felix Gausch, Toni Hofer, Mart Horn, Mitch Ingham, Wolfgang Kleissl, Ben Kuipers, Bernhard Rinner, Louise Travé-Massuyès, Franz Wotawa, and Feng Zhao

– thanks –

Graz, March 2005   *Michael W. Hofbaur*

Contents

Notation.. XV

2 Hybrid Estimation at a Glance 1
   2.1 Hybrid Model... 3
   2.2 Hybrid Estimation..................................... 6
       2.2.1 Full Hypothesis Hybrid Estimation 7
       2.2.2 Multiple-Model Estimation 10
       2.2.3 Focused Estimation............................... 11
       2.2.4 Unknown Mode 18
       2.2.5 Focused Real-Time Estimation of Complex Systems ... 21
   2.3 Hybrid Estimation in Automation 22
   2.4 Related Research 23
       2.4.1 Stochastic Systems and Kalman Filtering............. 23
       2.4.2 Hybrid Systems 23
       2.4.3 Multiple-Model Filtering 25
       2.4.4 Dynamic Programming and Best-First Search 26
       2.4.5 Qualitative Reasoning and Model-Based Diagnosis 26
       2.4.6 Fault Detection and Isolation (FDI) 26

3 Probabilistic Hybrid Automata 29
   3.1 Hidden Markov Models................................... 29
   3.2 Probabilistic Hybrid Automata........................... 32
   3.3 Concurrent Probabilistic Hybrid Automata 39
   3.4 PHA and cPHA Execution 41

4 Hybrid Estimation....................................... 45
   4.1 Traditional Estimation 45
   4.2 Hybrid Estimation (Non-switching Case) 49
   4.3 Full Hypothesis Hybrid Estimation 52
   4.4 Multiple-Model Estimation 66
   4.5 Focused Hybrid Estimation 70

4.5.1 Hybrid Estimation as Shortest Path Problem 72
4.5.2 Suboptimal Search Methods for Hybrid Estimation 94
4.6 Unknown Mode and Filter Decomposition 103
4.6.1 Unknown Mode 105
4.6.2 Filter Decomposition 106
4.6.3 Graph-Based Decomposition and Filter Cluster
Calculation 110
4.6.4 Filter Cluster Deduction for Hybrid Estimation 114

5 Case Studies 119
5.1 Three Component Example 119
5.2 Advanced Life Support System - BIO-Plex 125

6 Conclusion 139
6.1 Monograph Revisited 139
6.2 Future Work 140

References 143

Notation

Abbreviations

AI	Artificial Intelligence
BFSG	best-first successor generation
CDF	cumulative distribution function
c-HME	clustered hybrid (mode) estimation
cPHA	concurrent probabilistic hybrid automaton
DP	dynamic programming
FDI	fault detection and isolation
FFE	full fringe estimate
FMEA	failure mode effect analysis
GDE	General Diagnostic Engine
GPBn	generalized pseudo-Bayesian estimation
HME	hybrid (mode) estimation algorithm
HMM	hidden Markov Model
IMM	interacting multiple-model estimation
MAP	maximum a posterior
MLM	most likely mode
PDG	probability density function
PGC	plant growth chamber of the advanced life support system
PHA	probabilistic hybrid automaton
SCC	strongly connected component
SD	structural determined
SO	structural observable
SOD	structural observable and determined
vs-IMM	variable structure interacting multiple-model estimation

General Conventions

Meaning	Convention	Examples
scalars	lower case, italic	x_c, u_c
scalar function	lower case italic	$f(\cdot)$, $g(\cdot)$
vector, variable set	lower case, bold	$\mathbf{x}_c$, $\mathbf{u}_c$
vector valued function	lower case, bold	$\mathbf{f}(\cdot)$, $\mathbf{g}(\cdot)$
variable domain	upper case, calligraphic	$\mathcal{X}_d$, $\mathcal{U}_d$
sequence, set	upper case, italic	U, Y
matrix	upper case, bold	$\mathbf{A}$, $\mathbf{P}$
continuous valued variable	subscript c	$\mathbf{x}_c$, $\mathbf{u}_c$, u_c
discretely valued variable	subscript d	$\mathbf{x}_d$, x_d, $\mathbf{u}_d$
hybrid variable	no c/d subscript	$\mathbf{x}$, $\hat{\mathbf{x}}$
time index	superscript k or $,k$	$\mathbf{x}_k$, $\mathbf{x}_{c,k}$
automaton index	subscript j for vectors	$\mathbf{x}_{cj}$, $\mathbf{x}_{cj,k}$
vector component index	subscript i for scalars	x_{ci}, $x_{ci,k}$
hypothesis index	superscript (i)	$\hat{\mathbf{x}}_{c,k}^{(i)}$

List of Symbols

Symbol	Description	Page
$2^{\mathcal{S}}$	power set (set of all subsets) of a set $\mathcal{S}$	33
$\mathbf{A}$	system matrix of linear model	46
$a_{i,j}$	arc $n_i \xrightarrow{a_{i,j}} n_j$ of search tree	75
$\mathcal{A}$	probabilistic hybrid automaton	33
$\mathbf{B}$	input matrix of linear model	46
b, b_k	probability (mass) distribution for estimation hypotheses	51
$b_k^{(j)}$	probability (belief) for estimation hypothesis j	56
b_x^-, b_x^+	state bounds for continuous transition guard	34
b_u^-, b_u^+	input bounds for continuous transition guard	34
$\mathbf{C}$	output matrix of linear model	46
c_i	boolean guard for PHA transition specification	39
c_{ud}	propositional logic formula for transition guard	34
$\mathcal{CA}$	concurrent probabilistic hybrid automaton	39
$\mathcal{CG}$	causal graph	110
$\mathbf{D}$	direct transmission matrix of linear model	46

Symbol	Description	Page
$E\{\cdot\}$	expected value	46
$\mathbf{e}$, $\mathbf{e}_k$	estimation error	47
$F(\cdot)$	equation spec. function for an automaton	33
F_{DE}	set of discrete-time difference equations	33
F_{AE}	set of algebraic equations	33
$f(\cdot)$	utility function for best-first search	13, 80
$\mathbf{f}(\cdot)$	nonlinear function for difference equation	38
$\mathcal{F}_{DE}$	difference equation domain	33
$\mathcal{F}_{AE}$	algebraic equation domain	33
$g(\cdot)$	path cost for best-first search	13, 80
$\mathbf{g}(\cdot)$	nonlinear function for output equation	38
$h(\cdot)$	admissible heuristic for the path cost	13, 80
K, K_k	Kalman filter gain	48
κ	initial fringe size of N-step hybrid estimation	94
M_d, $M_{d,k}^{(j)}$	mode sequence	43
m_i, $\mathbf{m}_j$	automaton mode	33
μ_{ij}	IMM mixing probabilities	67
n_x	number of continuous state variables	33
n_u	number of continuous input variables	33
n_y	number of continuous output variables	33
n_v	number of continuous disturbance variables	33
n_i, n_j	nodes of a search tree	72, 75
η_k	number of estimates for trajectory hypotheses	100
$\mathcal{N}_k$	fringe nodes for leading trajectory hypotheses	92
$P_{\mathcal{O}}$	probabilistic observation function	30
$\bar{P}_{\mathcal{O}}$	modified probabilistic observation function	74
$P_{\mathcal{T}}$	probabilistic transition function	30
P_{Θ}	initial state spec. of HMM	31
$P(\cdot)$, $P(\cdot\|\cdot)$	probability, conditional probability	
$\bar{P}(\cdot)$	unnormalized likelihood	15
$p(\cdot)$	probability density function (PDF)	
p_c, $p_{c,k}^{(l)}$	continuous state estimate (distribution)	55
$\bar{p}_{c,k}^{(j)}$	mode conditioned estimates of IMM	68
$\mathbf{P}$	covariance matrix for the state estimate	46
$p_{\tau i}$, p_{τ}	probability specification PHA transition	33
$\Phi(\cdot)$	cumulative distribution function (CDF)	63

Symbol	Description	Page
$\mathbf{Q}$	covariance matrix of state disturbances	46
$q_{xc}(\cdot),\ q_{uc}(\cdot)$	nonlinear transition guard functions	34
$\mathbf{R}$	covariance matrix of output disturbances	46
$\mathbf{r},\ \mathbf{r}_k$	estimation innovation (prediction residual)	47
$r,\ r_i$	reset specification of a PHA transition	33
$\mathbb{R}$	set of real numbers	
$\mathbf{S}$	covariance matrix of the filter innovation	47
$\mathcal{SCC}$	strongly connected component	112
T_s	sampling time	4, 33
$T(\cdot)$	transition specification of a PHA	33
$\tau,\ \tau_i$	transition tuple of the PHA's transition	33
$\mathbf{u},\ \mathbf{u}_k$	hybrid input variables	33
$u_c,\ u_{ci,k}$	scalar continuous input variable	4, 33
$\mathbf{u}_c,\ \mathbf{u}_{cj,k}$	continuous input variables (vector)	33
$u_d,\ u_{di,k}$	scalar discrete input (command) variable	33
$\mathbf{u}_d,\ \mathbf{u}_{dj,k}$	discrete input (command) variables	33
$U,\ U_k$	hybrid input sequence up to time k	53
$U_c,\ U_{c,k}$	continuous input sequence (trace) up to time k	53
$U_d,\ U_{d,k}$	continuous output sequence (trace) up to time k	53
$\mathcal{U}_d$	domain for discrete input variables	33
$\mathbf{v}_c,\ \mathbf{v}_{cj,k}$	disturbance variables (vector)	33
$v_c,\ v_{ci,k}$	scalar disturbance variable	33
$\mathbf{v}_{cx},\ \mathbf{v}_{cj,k}$	state disturbance variables (vector)	46
$\mathbf{v}_{cy},\ \mathbf{v}_{cj,k}$	output disturbance variables (vector)	46
$\mathbf{w},\ \mathbf{w}_k$	hybrid I/O variables (vector)	33
$\mathbf{w}_c,\ \mathbf{w}_{cj,k}$	continuous I/O variables (vector)	33
$w_c,\ w_{ci,k}$	scalar continuous I/O variable	33
$\mathbf{w}_d,\ \mathbf{w}_{dj,k}$	discrete I/O variables (vector)	33

Symbol	Description	Page
$\mathbf{x}$, $\mathbf{x}_k$	hybrid state variables (vector)	33
x_c, $x_{ci,k}$	scalar continuous state variable	4, 33
$\mathbf{x}_c$, $\mathbf{x}_{cj,k}$	continuous state variables (vector)	39
$\mathbf{x}'_{c,k}$	continuous state immediately after mode transition	39
x_d, $x_{di,k}$	scalar discrete state (mode)	33
$\mathbf{x}_d$, $\mathbf{x}_{dj,k}$	discrete state (mode) variables (vector)	39
$\hat{\mathbf{x}}$, $\hat{\mathbf{x}}_k^{(l)}$	hybrid state estimate	46, 53
$\hat{\mathbf{x}}_c$, $\hat{\mathbf{x}}_{cj,k}^{(l)}$	continuous state variables (vector) estimate	46, 53
$\hat{\mathbf{x}}_d$, $\hat{\mathbf{x}}_{dj,k}^{(l)}$	discrete state (mode) variables (vector) estimate	46, 53
X, X_k	discrete time trajectory	42
$\hat{X}$, $\hat{X}_k^{(l)}$	discrete time trajectory estimate	53
X_0	PHA initial state specification	33
$\mathcal{X}_k$	fringe estimates at k	92
$\mathcal{X}_d$	domain for discrete state (mode)	33
y_c, $y_{ci,k}$	scalar continuous output variable	4, 33
$\mathbf{y}_c$, $\mathbf{y}_{cj,k}$	continuous output variables (vector)	39
Y_c, $Y_{cj,k}$	continuous output sequence (trace)	42
$\mathcal{Y}_d$	domain for discrete output (of HMM)	31
$\mathbf{z}_c$, $\mathbf{z}_{cj}$	continuous variables (vector)	36
Z, Z_k	hybrid input/output sequence	56
ζ	number of component automata of a cPHA	37

Hybrid Estimation at a Glance

One of our test applications is a simulation study of the BIO-Plex Test Complex at NASA Johnson Space Center, a five chamber facility for evaluating biological and physiochemical Martian life support technologies. It is an artificial, biosphere-type, closed environment, which must robustly provide all the air, water, and most of the food for a crew of four without interruption. Plants are grown in two plant growth chambers where they provide the food for the crew and convert the exhaled CO_2 into O_2. In order to maintain the closed-loop system, it is essential to control the resource exchange between the plant growth chambers and the other chambers without endangering the crew. For the scope of this book, we restrict our evaluation to CO_2 and O_2 control in one plant growth chamber (PGC) as shown in Fig. 2.1. This subsystem is composed of several components, such as redundant flow regulators that provide continuous CO_2 supply, redundant pulse injection valves that provide a means for increasing the CO_2 concentration rapidly, a light system, redundant O_2 concentrators and the plant growth chamber (PGC) itself.

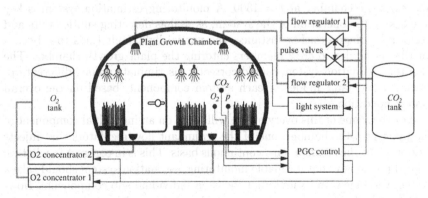

Fig. 2.1. BIO-Plex plant growth chamber subsystem.

The control system maintains a plant growth optimal CO_2 concentration of 1200 ppm and an ambient oxygen level of 21 vol.% during the day phase of the system (20 hours/day). The relatively high CO_2 level is unsuitable for humans, hence the gas concentration is lowered to 500 ppm whenever crew members request to enter the chamber for harvesting, re-planting, or other service activities. Safety regulations require that the system inhibits the high-volume gas injection via the pulse-injection path while crew members are in the plant growth chamber. Sensors are available to record the entry and exit of crew members. However, sensors are known to fail and a capable monitoring/estimation system has to robustly estimate the current operational mode and health state of the system. This involves determining the mode of each component (for example, flow regulators, sensors, valves, concentrators), and detecting the operational situation for the overall system (for example, plant growth mode, service mode with or without presence of crew members in the chamber, etc.) based on noisy measurements of various physical entities (for example, CO_2 flow rate, CO_2 and O_2 gas concentration, chamber pressure, etc.). Figure 2.2 shows a typical trace of the CO_2 gas concentration during operation (one time-step represents the duration of one minute). The crew requests to enter the plant growth chamber at the time-point $t = 600$. After lowering the concentration to the safe level of 500 ppm, the system unlocks the door and the crew enters at $t = 850$. Their entry can be seen from the slight disturbance of the CO_2 gas concentration that is due to the additional exhaled CO_2. The control mechanism adapts the gas injection to a new value so that the overall concentration is maintained at the 500 ppm level after a short adaption phase. A light fault happens at $t = 1000$. The reduced illumination of the chamber leads to a lower photosynthesis activity of the plants, so that less CO_2 is consumed. This causes the imbalance that again is corrected by the chamber control system. The crew repairs the fault at $t = 1100$ and exits the chamber at $t = 1300$. A monitoring/estimation system is key to tracking the operation of the system, as well as detecting subtle faults and performing diagnoses. For instance, the partial light fault leads to a behavior that is similar to crew members entering the plant growth chamber. The monitor/estimation system should correctly discriminate among different operational and fault modes of each system component, based on the overall system operation.

For the scope of this overview we will focus on an individual component of the plant growth chamber, one of the redundant flow regulators that injects CO_2 gas into the chamber on a continuous basis. This is in contrast to what we argued previously that a capable monitoring/estimation system should take a system-wide view and should not focus on individual sub-components. However, it will allow us to keep the presentation concise, without the requirement to add too much detail that could mask the overall story. The generalization to estimation that takes a system-wide view is obtained automatically when moving from the single component to concurrently operating components. We shall see later that the overall estimation scheme decomposes the system into

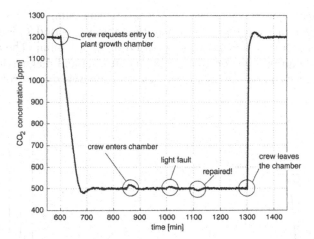

Fig. 2.2. Typical trace of the CO_2 gas concentration during operation.

component clusters, whenever possible. However, it also maintains estimation hypotheses where large grain component-clusters, or even the system as a whole, are taken into account. This enables our hybrid estimation scheme to obtain additional information based on the system-wide interaction of the components. For example, whenever the flow sensor of a flow regulator fails, we can still estimate the in-flow due to its interaction within the plant growth chamber. For the following sections, however, we restrict the description to a single component and indicate the difficulties that arise with multi-component systems.

2.1 Hybrid Model

The flow regulator represents a low level component cluster that consists of a continuously actuated valve, a flow meter, and a low-level controller that adapts the valve's opening in order to provide a continuous gas-flow at the requested level. For the purpose of supervisory control, we can view these low-level components to comprise one generic flow regulator that utilizes a control mechanism which operates on a significantly faster time-scale than the supervisory control system. Figure 2.3a shows the response of the actuated gas flow rate based on the input signal. Whenever one is interested in determining the low-level health state of the flow regulator, one would choose a high sampling rate, so that the transition process can be judged, for example, to determine whether wear and tear lead to a degraded device with slower response and off-set errors. Contrariwise, from the perspective of supervisory control, one might view the transition phase as instantaneous and only judge the overall operation of the flow regulator. For instance, whether a commanded flow change was actuated correctly up to the time-point when the supervisory system takes the next measurement sample. This sampling strategy is illustrated in Fig. 2.3a, where the sampling period T_s is set to the supervisory control system's sampling period of $T_s = 1$ [min]. The appropriate level of detail depends on the specific monitoring and diagnostic task. In the following we choose the high level view of a supervisory control system.

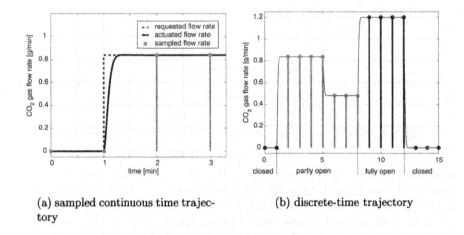

(a) sampled continuous time trajectory

(b) discrete-time trajectory

Fig. 2.3. Trajectories for the flow regulator.

We distinguish three operational modes to take the non-linear effect of cutoff and saturation into account, as indicated in Fig. 2.3b. Instead of using one non-linear model to describe the complex dynamic behavior, we will use specific dynamic models for the operational conditions. A *hybrid model* combines the continuous dynamic models with a discrete automaton-like model that captures the transitions among the operational *modes*.

The three distinct operational modes of our flow regulator are `closed`, `partly-open`, and `fully-open`. The modes constitute different constraints (equations) among the continuous input u_c, the actuated flow x_c and the noisy measurement y_c. We use stochastic difference equations to capture the model of the dynamic processes as follows:

$$\texttt{closed}: \quad x_{c,k} = 0$$
$$y_{c,k} = x_{c,k} + v_{c,k}$$

$$\texttt{partly-open}: x_{c,k} = u_{c,k-1}$$
$$y_{c,k} = x_{c,k} + v_{c,k} \tag{2.1}$$

$$\texttt{fully-open}: \quad x_{c,k} = x_{c,\max}$$
$$y_{c,k} = x_{c,k} + v_{c,k}\,.$$

All difference equations operate with a *sampling period* T_s and the sample-point k represents the time-point $t = kT_s + t_0$, where t_0 denotes the initial time-point. In the following we assume that the initial time-point is set to zero ($t_0 = 0$) so that $x_{c,k} = x_c(kT_s)$. The models capture the facts that there is no output flow in a closed valve, and an outflow with the maximal flow rate at a fully open valve[1]. For the case of a partly open valve, the flow regulator actuates the requested flow $u_{c,k}$ at the next time sample $k + 1$. As in reality, we assume that measurements are subject to noise and we model the measurement process in terms of additive noise v_c in all modes.

The overall operation of the flow regulator is captured by defining the conditional transitions among the modes. They are described best visually in a *transition graph* as shown in Fig. 2.4. Each node in the graph represents an operational mode of the system and the arcs represent possible transitions among the modes. The arcs are labeled with conditions that enable the associated transitions. For example, whenever the system is in the mode `closed` at time-point k and it experiences a continuous input value $u_{c,k} > 0$, then it will execute the transition `closed` $\rightarrow$ `partly-open` and proceeds at the mode `partly-open` up to the next time-point $k + 1$. The transition and the consecutive continuous evolution ensures that the flow regulator actuates the requested flow rate at the time-point $k + 1$, as defined by (2.1). Thus, the adaption of the flow takes place in between the two sampling-points k and $k + 1$.

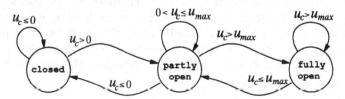

Fig. 2.4. (Nominal) transition graph of flow regulator.

It is our intend to model the system stochastically. This involves the continuous dynamics as indicated above, but also the automaton part of the model. For example, one might want to express the possibility of a stuck-closed valve, so that an actuation of the flow regulator leads to a partly open valve in most cases, but there also is a less likely chance that the valve remains closed. We obtain such a specification by using a *stochastic automaton model* that defines transitions probabilistically as shown in Fig. 2.5. Whenever the system is in mode `closed` at the time-point k and it experiences a continuous input value $u_{c,k} > 0$, then it will execute the nominal transition `closed` $\rightarrow$ `partly-open` with probability 0.9 or remains closed with probabil-

---

[1] For simplicity, we assume a constant pressure difference between the PGC and the CO_2 tank that feeds the flow regulator.

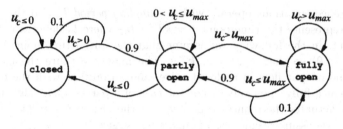

Fig. 2.5. Probabilistic transition graph of flow regulator.

ity 0.1. The latter transition *thread* captures the stuck-closed condition. This
illustrates our probabilistic transition model that specifies conditional transitions that *thread* over possible transition goal modes. A similar transition
model, but without guards, can be found in *Hidden Markov Models (HMM)*
[36, 86]. We utilize this model as the automaton basis, extend the transition
scheme with guarded transitions, and incorporate stochastic discrete-time dynamic models. This leads to a stochastic hybrid model that we call *probabilistic
hybrid automaton* [55], or *PHA* for short.

The model given above does not introduce specific fault modes that capture dedicated fault situations, such as stuck closed or stuck open valves.
Mostly, because the continuous behavior of the mode `closed` (or `fully-open`)
is identical with the behavior that is experienced in the case of the stuck
closed (or stuck open) situation. In general, however, it is possible to include
dedicated fault modes as we shall see later. Furthermore, we also will introduce another specific fault mode, the *unknown mode*. This mode captures
all unmodeled situations, such as a drift fault, offset fault, stuck at partially
open position etc. . For the moment, however, we restrict our description of
the flow regulator to a *hybrid model* that incorporates the 3 modes (`closed`,
`partly-open`, `fully-open`) only.

2.2 Hybrid Estimation

It is important that a monitoring and diagnosis system is able to accurately
estimate the *hybrid state* of a system in order to track the system's operation
and to detect the onset of subtle failures. The hybrid state x_k is comprised of
the continuous state $x_{c,k}$ and the mode $x_{d,k}$ of the system at a specific time-
point k. A hybrid estimator utilizes the noisy measurements and provides
both, an estimate for the continuous state $x_{c,k}$, as well as an estimate for
the mode $x_{d,k}$. An optimal hybrid estimation algorithm has to consider *every
possible evolution* of the system. This leads to the *full hypothesis estimator* for
hybrid systems. In the following we will introduce the full hypothesis estimation algorithm, demonstrate that it is computationally infeasible, and show
various mechanisms for sub-optimal hybrid estimation, such as our proposed
hybrid estimation scheme [55].

2.2.1 Full Hypothesis Hybrid Estimation

Let us assume that the flow regulator is closed initially ($k = 0, t = t_0$). In terms of the model we specify this by the mode $x_{d,0} = $ closed and the continuous state $x_{c,0} = 0$. At the same time-point, we actuate the flow regulator with the continuous input signal $u_{c,0} = 0.5$. This input signal triggers a transition with the likely outcome partly-open and the less likely outcome closed. We take the hybrid systems view and assume that this transition takes place instantly at $t = t_0$ and requires an infinitesimally short period of time $[t_0 \; t_0 + \epsilon]$. Since we do not observe the transition directly, we have to consider both transition hypotheses and the consecutive evolution of the flow rate up to the next sample $k = 1$, which represents the time-point $t = 1T_s$. The hybrid estimator traces these possible evolutions. The transition probabilities P_T for the two possible transitions provide a *prior rating* for the two hypotheses. The probabilistic transition model of the flow regulator (Fig. 2.5), and the fact $u_{c,0} > 0$ provides the probability 0.9 for the hypothesis with the mode partly-open and 0.1 for the hypothesis with the mode closed. Figure 2.6 records this fact in terms of a *hypothesis tree*. The hybrid estimator also uses the initial state information for the continuous state and the actuated continuous input to provide a prior estimate for the continuous state, one estimate for the mode partly-open and one estimate for the mode closed. It calculates the one-step ahead prediction for both hypotheses according to the appropriate stochastic model (2.1) and obtains a continuous estimate in the form of a distribution (probability density function) among the continuous state space.

Once time proceeded up to $t_1 = t_0 + 1T_s$, we can utilize the new measurement $y_{c,1}$ and compare this value with the two predictions that can be drawn from the continuous estimates of both hypotheses. This comparison provides two things: firstly, we obtain a *measure of likelihood* $0 \le P_O \le 1$ that expresses how well an estimate agrees with the observation, and secondly, we can use the estimate to refine the prediction in the sense of the prediction/correction

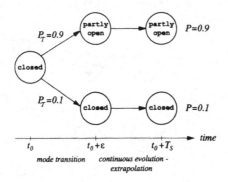

Fig. 2.6. One-step hypotheses tree based on prior information.

scheme, such as it is done in a Kalman Filter. For our example we assume that the measurement at t_1 provides the value $P_O = 0.8$ for the estimate at mode `partly-open`, and $P_O = 0.1$ for the estimate at mode `closed`. This leads to the posterior estimates $\hat{x}_{c,1}^{(1)}$ and $\hat{x}_{c,1}^{(2)}$ for the continuous states and an updated posterior probabilities $P_1^{(1)}$ and $P_1^{(2)}$ for each hypothesis at time-step 1. This is indicated in Fig. 2.7.

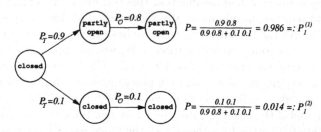

Fig. 2.7. One-step hypotheses tree based on posterior information.

The hybrid estimator can selectively repeat this process as time proceeds. For every estimate at the time-point $t = (k-1)T_s$, it considers the possible transitions and estimates the resulting trajectory up to the consecutive time-point $t = kT_s$. This process can be seen as building a *full hypothesis tree* for time-step k. The full hypothesis tree encodes the overall hybrid estimate for the system up to the final time-step k based on the prior information (initial state $x_{c,0} = 0$, $x_{d,0} = $ `closed`), the input sequence $\{u_{c,0}, u_{c,1}, \ldots, u_{c,k-1}, u_{c,k}\}$ and the measurements $\{y_{c,1}, y_{c,2}, \ldots, y_{c,k-1}, y_{c,k}\}$. Figure 2.8 visualizes the hy-

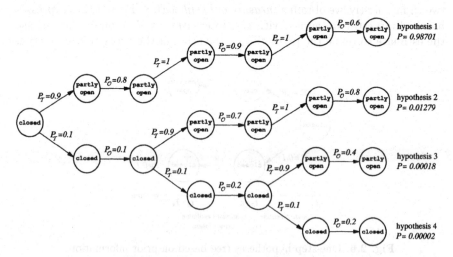

Fig. 2.8. Three-step hybrid estimation.

pothesis tree for the flow regulator at $k = 3$. Hypothesis 1 represents the most likely trajectory hypothesis for the flow regulator. It captures a trajectory with the mode sequence

$$\texttt{closed} \rightarrow \texttt{partly-open} \rightarrow \texttt{partly-open} \rightarrow \texttt{partly-open}$$

and the fringe estimate

$$\hat{x}_3^{(1)} = \langle \texttt{partly-open}, \hat{x}_{c,3}^{(1)} \rangle$$

that quantifies the hybrid state of the particular trajectory at time-step $k = 3$. In this section we use the notation $\hat{x}_{c,\kappa}^{(i)}$ to denote the continuous state estimate of hypothesis i at time-step κ. This is a slight abuse in notation since stochastic estimators provide a multivariate probability density function $p_{c,\kappa}^{(i)}$ with the mean $\hat{x}_{c,\kappa}^{(i)}$. However, for the sake of clarity in this overview section we do not make this distinction and simply use the mean's symbol to denote the continuous estimate.

The estimation algorithm not only provides estimates for the possible hybrid state trajectories, it also provides their likelihood. For example, 0.98701 for the hypothesis 1 with the fringe state estimate $\hat{x}_3^{(1)}$. These conditional probabilities specify the weighting of the particular hypothesis for the overall hybrid estimate. For example, one would obtain the *overall* continuous estimate $\hat{x}_{c,k}$ at $k = 3$ by *merging* the 4 fringe estimates according to their likelihood

$$\hat{x}_{c,k} = 0.98701\hat{x}_{c,k}^{(1)} + 0.01279\hat{x}_{c,k}^{(2)} + 0.00018\hat{x}_{c,k}^{(3)} + 0.00002\hat{x}_{c,k}^{(3)} \, .$$

The hypotheses as a whole also define the likelihoods for the individual modes at the time-step $k = 3$. These probability values sum-up the probabilities of the individual modes as shown in the Table 2.1.

Table 2.1. Posterior mode probabilities at time-step 3.

mode	hypotheses	probability
partly-open	{1, 2, 3}	0.99998
closed	4	0.00002
fully-open	-	0.00000

The continuous estimate at the most likely mode **partly-open**

$$\hat{x}_{c,3}|_{x_{d,3}=\texttt{partly-open}} = \frac{1}{0.99998}(0.98701\hat{x}_{c,k}^{(1)} + 0.01279\hat{x}_{c,k}^{(2)} + 0.00018\hat{x}_{c,k}^{(3)})$$

represents the *mode-conditioned state estimate* for the time-step $k = 3$ with the likelihood $0.98701 + 0.01279 + 0.00018 = 0.99998$.

Table 2.2. Example posterior mode probabilities at time-step k.

hypothesis	mode	probability
1	partly-open	0.40
2	closed	0.30
3	closed	0.15
4	fully-open	0.05
$\vdots$	$\vdots$	$\vdots$

Depending on whether we are interested in the most likely trajectory or in the hybrid estimate at a specific time-step, we have to interpret the full hypothesis tree accordingly. Focusing on the most likely hypothesis only might lead to an incomplete or even incorrect overall estimation result. For example, consider a different set of estimates shown in the Table 2.2. Taking only the most likely hypothesis into account, one would come to the wrong conclusion that the most likely mode is partly-open (with likelihood 0.4). The hypotheses 2 and 3 describe estimates for mode closed. Together, these hypotheses provide a larger portion in the probability space (45%), than hypothesis 1 for the mode partly-open. Considering the first 4 hypotheses only, one would conclude that the most likely mode is closed (likelihood 0.45), followed by the mode partly-open (likelihood 0.40), and fully-open (likelihood 0.05). However, the four hypotheses only specify 90% of the probability space and the remaining omitted hypotheses can still change the mode estimation result. This is due to the fact that their probabilities sum up to 0.10, which is larger than the difference of 0.05 between the likelihoods for the leading modes closed and partly-open. This hypothetical example should demonstrate that we cannot easily omit the trajectory hypotheses with low likelihood. One should either calculate the full hypothesis tree or carefully omit hypotheses with low likelihood in order to obtain a correct estimation result.

Full hypothesis estimation, however, involves a number of trajectory hypotheses that is (worst case) exponential in the number of time-steps considered. Thus, tracking all possible trajectories of a system is almost always intractable because the number of trajectory hypotheses becomes too large after only a few time-steps. As a consequence, it is inevitable to use *approximate* hybrid estimation schemes that merge trajectory hypotheses, and/or prune unlikely hypotheses, so that the number of hypotheses under consideration stays within a certain limit.

2.2.2 Multiple-Model Estimation

A simple approach to cope with the exponential explosion of trajectory hypotheses is to combine them and consider different mode sequences only for the last n estimation-steps. This is the essence of the generalized pseudo-Bayesian

(GPBn) approaches [2] of a family of so-called *multiple-model estimation* algorithms. The first-order version GPB1, for instance, merges all hypotheses after building the one-step hypothesis tree and uses this overall hybrid estimate as initial state estimate for the next time-step. In general, an nth-order version GBPn considers all possible mode sequences in the last n estimation-steps and merges them into one single hybrid state estimate. However, these algorithms come with the price that they require at most l^n concurrent filters, where l denotes the number of modes.

A good trade-off between computational cost and estimation quality is achieved by the *interacting multiple-model (IMM)* algorithm [20]. IMM provides an estimate with quality of the GPB2 algorithm, but requires at most l concurrent filters (one filter per mode of the hybrid model). Each filter uses a different combination (mixing) of the mode-conditioned estimates as the initial value for the estimation at the next time-step $k + 1$.

Multiple-model estimation algorithms maintain a certain (fixed or adaptive) number of filter results, and merge their results according to the mixing-scheme of the particular multiple-model algorithm. The methods work fine whenever the number of modes stays within some reasonable bound so that a concurrent operation of the filters is conform with the time and space requirements of real-time operation. However, they do not scale well with the number of modes and cannot be applied whenever the number of modes becomes very large (in the order of thousands or more). Our intended applications, however, fall into this category of complex systems with a vast number of modes, as a consequence, we apply a different sub-optimal estimation scheme that carefully explores the full hypothesis tree and that focuses onto the most likely hypotheses.

2.2.3 Focused Estimation

Given the flow regulator example above, it is evident that few hypotheses take up the major portion of the probability space. For example, after the first step, hypothesis 1, which describes the flow regulator at its partly open position, has the likelihood 0.986, thus it takes the major portion of 98.6% of the probability space. A similar observation can be made for systems with a significantly larger number of modes. In those systems a small fraction of possible hypotheses typically covers 99% of the possible outcomes [104]. Discrete estimation methods, such as the General Diagnostic Engine (GDE) [29] or the Livingstone and Titan systems [104, 101], that build upon the model-based reasoning paradigm exploit this fact successfully. Livingstone, for instance, was successfully demonstrated on-board of the DS-1 space probe, a system with approximately 10^{48} modes of operation [105]. Key to the capability of handling such a large number of modes is to carefully *enumerate* possible hypotheses so that the estimation focuses onto the major set of possible hypotheses, only.

Hybrid Estimation as Search

The key for an efficient enumeration scheme is to formulate hybrid estimation as *search problem* that consecutively returns hypotheses, starting with
the leading one. Let us reconsider the hypothesis tree for the flow regulator
shown in Fig. 2.8. Every arc in the tree is associated with either a transition
probability P_T or the value of the observation function P_O. The resulting
likelihood of a hypothesis is obtained by taking the product of these numbers
along the path from the root node (initial state x_0) to one of the leaf nodes
(hybrid estimate $\hat{x}_k^{(i)}$) divided by a common scaling factor $c > 0$. Since the
scaling factor is the same for all leaf nodes, it is possible to reformulate the
hybrid estimation problem as a *shortest path* problem, where each arc in the
tree is associated with a path length, or *cost*, according to its transition probability P_{Ti}, or the observation functions P_{Oi}, respectively. Search seeks for the
shortest path, that is, the path with the lowest cost. This path specifies the
leading trajectory hypothesis for hybrid estimation. In shortest path problems
the cost of arcs along the path are combined using addition. Our probabilistic
framework, however, uses multiplication and we seek for the largest conditional probability. Therefore, we use the the standard approach of taking the
negative logarithm of P_i as the cost for the individual arcs in the tree:

$$\text{cost}_i := -\ln(P_i) \, ,$$

where P_i denotes either a transition probability P_{Ti} or an observation function
P_{Oi}, depending on the type of the arc. Figure 2.9 shows the 3-step hypothesis
tree of Fig. 2.8 with cost labeling and a numbering scheme for the nodes in
the tree.

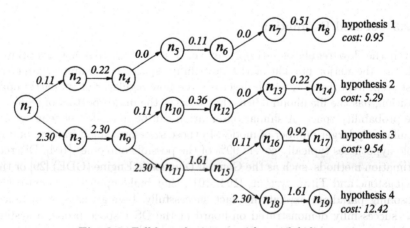

Fig. 2.9. Full hypothesis tree with cost labeling.

A wide verity of algorithms exist for shortest path problems. Dynamic
programming [16, 19] is probably the best known solution method. Dynamic

Programming (DP) is a good choice whenever one solves a problem that is expressed in terms of a *graph*, with redundant paths to nodes within the graph. DP applies *Bellman's principle of optimality* and keeps only the best path to a node, whilst exploring the graph. Hybrid estimation, however, maps to a hypothesis *tree* that has, per definition, only a unique path to each node within the tree. As a consequence, applying DP to our problem does not provide any advantage over performing the full hypothesis estimation directly. Therefore, we apply an alternative search methodology that falls into the class of *best-first search* algorithms.

The search algorithm starts at the root node n_1, which represents the initial state, and carefully expands the tree toward the leaf nodes. For this purpose, best-first search incrementally expands search tree nodes that seem most promising. The 'promise' of a node n_i is measured by *utility* $f(n_i)$, which combines the cost $g(n_i)$ from the root node to the node n_i with the (conservative) estimate $h(n_i)$ of the cost to go. Unexpanded nodes are kept in an ordered list (search agenda) so that the node with the best utility is expanded first. This motivates the name best-first search. The property of the search algorithm highly depends on the evaluation function $f(\cdot)$ that determines the utility of a node, and as a consequence the path that the search algorithm explores first. We apply a variant of best-first search that is known as A* in literature. This algorithm utilizes an evaluation function

$$f(n_i) = g(n_i) + h(n_i) \,,$$

with an *admissible heuristic* $h(n_i)$ that *never overestimates* the cost to go. This ensures that the search procedure not only focuses onto the leading trajectory hypotheses, but it also provides the trajectory estimates consecutively in the order of decreasing likelihood [49].

Consecutively generating the estimates for trajectory hypotheses in the order of decreasing likelihood enables us to terminate the search procedure any-time, whenever we run short of computational resources. Any continuation would only add additional less likely estimates. This property is very helpful for the application within a real-time environment and is known as *any-time/any-space* in literature since we can terminate hybrid estimation whenever we run out of computation time (any-time) or whenever we exceed memory-space constraints (any-space). In this way, we achieve a good trade-off between the limited computational resources of a real-time system and the estimation accuracy.

Let us demonstrate the A* strategy with the hypothesis tree of Fig. 2.9. The search algorithm expands the root node n_1 that encodes the initial state of the hybrid estimation problem in the first step. The path from the initial state to node n_2 is determined by the transition from closed → partly-open and has an assigned probability of 0.9, thus the cost to go for the first arc is $g(n_2) = -\ln(0.9) = 0.11$. The 'transition' closed → closed is less likely (0.1) and induces a higher cost of $g(n_3) = -\ln(0.1) = 2.30$ for the corresponding

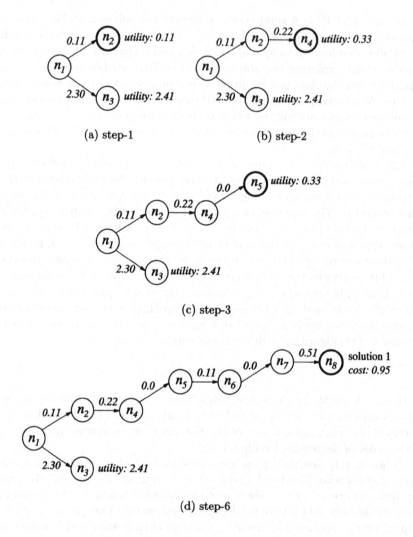

(a) step-1   (b) step-2

(c) step-3

(d) step-6

Fig. 2.10. Best-first search, step 1 to 6.

arc. It is important that the evaluation of the heuristic is computationally efficient. The computationally intensive operations of hybrid estimation are the filtering step and the evaluation of transition guards. Both operations are to be avoided for estimating the cost to go. Therefore, we assume the best possible filtering result that provides a perfect match of estimate and observation ($P_{Oi} = 1$) and evaluate possible transitions based on the transition threads only. Of course, this will lead to a conservative guess, however, in taking this assumption, we obtain an computationally efficient method that retains the admissibility of the heuristic. The values for the heuristic can be deduced as follows: The conservative estimate for the observation functions P_O of the following arcs along the path to the leaf node is $P_O = 1$, as indicated above. This assigns an estimated cost of 0 to all arcs that represent the estimation-steps. With respect to possible transitions from the mode partly-open, we consider the sequence of transitions with highest probability, given the transition guards are satisfied. Since the mode partly-open has a self loop with probability $P_T = 1$ it is easy to see that the sequence of transitions with highest probability is simply to stay at mode partly-open, thus the heuristic for node n_2 is $h(n_2) = 0$ inducing a utility $f(n_2) = 0.11$ that is simply the cost to the node. The heuristic value for node n_3 can be deduced similarly. Again we take the default cost of 0 for observation arcs. The best continuation in terms of possible transitions from mode closed onward, is to take the transition closed $\rightarrow$ partly-open with cost 0.11, and then to remain at partly-open with cost 0. This implies a value for the heuristic function of $h(n_3) = 0.11$ and leads to the utility of node n_3:

$$f(n_3) = 2.30 + 0.11 = 2.41 .$$

Given the utility of the nodes n_2 and n_3, A* continues with the expansion of n_2. This expansion leads to a new node n_4 with utility 0.33 (Fig. 2.10b) that is then expanded further, because its utility is smaller than the utility of the node n_3. The recursive process proceeds (Figs. 2.10a to 2.10d) until it finds the least-cost path from node n_1 to node n_8. This path in the hypothesis tree represents the most likely trajectory estimate for $k = 3$ (hypothesis 1 in Fig. 2.8). Whenever time or memory space is short already, we could terminate here and provide the most likely estimate as estimation result, otherwise, we can continue to search for the next best path by continuing the search process starting at the partially expanded hypothesis tree of Fig. 2.10d. The continuation up to the next solution is shown in Fig. 2.11. This provides the estimation hypothesis 2 in the form of a path with cost 5.29.

In a real-time environment we do not have the resources to calculate the full hypothesis tree, in general. Therefore, we need some criteria to decide, whether the hypotheses found so far represent a sufficiently good approximation for the current hybrid state. The path costs are directly related to the unnormalized likelihoods $\bar{P}$ of the hypotheses

$$\bar{P}(\text{hypothesis}) = e^{-\text{path\_cost(hypothesis)}} .$$

Furthermore, we can assess the maximum number of possible hypotheses at time-step k, based on the model's possible transitions. As in the case of the heuristic it is important that we determine this number efficiently, thus we do not evaluate the transition guards, but determine a *conservative upper bound* of the number of possible hypotheses given the transition threads only. In our case, this would mean that the upper bound on hypothesis for a hypothesis

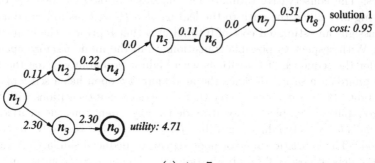

(a) step-7

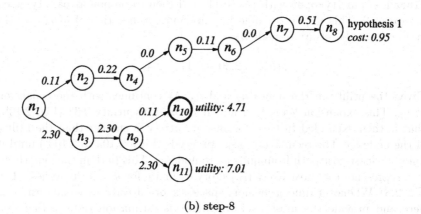

(b) step-8

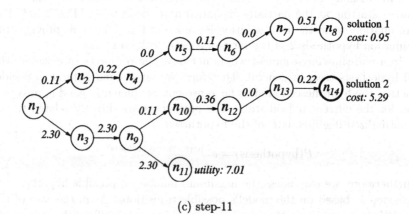

(c) step-11

Fig. 2.11. Best-first search, step 7 to 11.

tree that starts at the mode `closed` is $2 \times 3 \times 3 = 18$ hypotheses[2]. The unnormalized likelihoods of the first two hypotheses are given by

$$\bar{P}_1 = 0.387, \quad \bar{P}_2 = 0.005 \, .$$

For the remaining hypotheses, we know that their unnormalized likelihoods are smaller or at most equal to $\bar{P}_2 = 0.005$. Again, taking this upper bound as conservative guess, we obtain the upper bound for the normalization factor of

$$c_{max} = \bar{P}_1 + \bar{P}_2 + (18 - 2)\bar{P}_2 = 0.472 \, .$$

With this value, we can estimate a lower bound for the likelihood of hypothesis 1

$$P_1 \geq \frac{\bar{P}_1}{c_{max}} = \frac{0.387}{0.472} = 0.820 \, .$$

The first two hypotheses take up at least

$$\frac{\bar{P}_1 + \bar{P}_2}{c_{max}} 100 = 83.1\%$$

of the overall probability space. This number can serve as an indicator about how well the set of hypotheses that were calculated so far approximate the overall hybrid estimate for the time-step under consideration.

The first two hypotheses define estimates at mode `partly-open`. It is now interesting to see, whether we can guarantee that the most likely mode at $k = 3$ is in fact `partly-open`, based on the two leading hypotheses only. Both estimates together provide an unnormalized likelihood of $0.387 + 0.005 = 0.392$ for the mode `partly-open`. Based on our conservative estimate for the number of possible hypothesis at $k = 3$, we can determine that the remaining hypotheses add up to a value of at most

$$(18 - 2)\bar{P}_2 = 0.08 \, .$$

Thus, even if all other hypotheses are at a different mode than `partly-open`, they can never revert the decision on the most likely mode anymore. Thus, in calculating a portion of the hypothesis tree only, we still guarantee that the mode estimate is correct. We will later use this criteria do decide on the number of hypotheses that we need in order to guarantee correctness of our estimation result.

A$*$ search together with a termination criteria as given above allows us to perform sub-optimal hybrid estimation where only few hypotheses are used to represent the hybrid estimate at a given time-step. The benefits of this approach might not be that obvious, given the simple flow regulator component. It would still be possible and also reasonable to utilize an alternative

---

[2] Mode `closed` has at most two successors, whereas an arbitrary mode has at most 3 successors. Therefore, the mode `closed` has at most two successors after the first step, each of which can have at most 3 successors in the second step etc. .

multiple-model estimation algorithm for it. However, the number of modes in more complex systems easily overwhelms the computational requirements for multiple-model estimation so that one has to focus on a small set of likely hypotheses only. A∗ search highly focuses onto few hypotheses, compared to the overall number of modes. However, it will still cause the hypotheses tree to steadily grow, as the estimation time proceeds. Even if the growth is much smaller than the growth of the full hypothesis tree, it is still impractical for a real-time implementation that executes over a long period of time.

We solve this problem by growing the tree for N time-steps only. Whenever the time-index k exceeds this value, we will restart the search from the leading set of κ hypotheses at the time-step $k - N$. Of course, we will reuse the previously grown tree as much as possible and discard only those parts of the tree that are neglected. Figure 2.12 illustrates this strategy for $N = 1$. At each time-step $k - 1$, we take the κ leading hypotheses and perform the A∗ search to deduce the leading set of successor hypotheses at the time-step k. Although the A∗ search technique is guaranteed to provide the leading set of hypotheses, we obtain an approximative hybrid estimation scheme, since we start the search from a limited number of hypotheses and neglect the less likely ones. One can interpret the overall strategy as a two level search problem that utilizes A∗ as its low-level strategy to perform the enumeration of successor hypotheses, and a high level *beam search* strategy that restricts the number of hypotheses to the set of κ most likely ones. The level of approximation can be tailored to the complexity of the estimation problem, as well as the computational resources by adapting the beam-size κ and the tree-depth N.

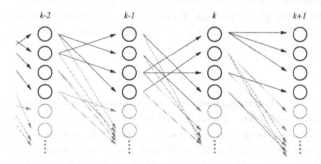

Fig. 2.12. Beam-search tree for hybrid estimation.

2.2.4 Unknown Mode

The mode estimation scheme as defined above, as well as the standard multiple-model estimation algorithms, assume that the system exhibits a mode of operation within the set of modes that is captured by the model. The models, however, represent approximations of the real world and many recent efforts in estimation and fault detection and isolation (FDI) were devoted to building *robust* estimation/diagnosis algorithms that can cope with unavoidable inaccuracy and incompleteness of the model [39, 26]. Nevertheless, unanticipated failures do occur in real world systems and robust estimation and diagnosis methods either fail to detect such an unusual operational condition (their robustness leads to a wrong classification) or perform in an unexpected way. As a consequence, it is desirable to extend the estimation and diagnosis capability so that it can cope with and identify an *unknown mode* of operation.

Another benefit of including an unknown mode is to capture a verity of failure modes that are difficult to model. For example, consider the flow regulator again. Potential fault modes are behaviors that exhibit a constant or a slowly varying offset or drift of the actuated flow rate. The magnitudes of possible offset-levels or drift-rates, as well as their directions, are difficult to predict a priori. As a consequence, it is hard to define dedicated models that capture these faults sufficiently well.

The concept of the *unknown mode* is central to discrete model-based diagnosis [46]. Its underlying concept of constraint suspension [28] allows diagnosis of systems where no assumption is made about the behavior of one or several components of the system. In this way, model-based diagnosis schemes, such as the *General Diagnostic Engine (GDE)* [29] or *Sherlock* [30], capture unspecified and unforeseen behaviors of the system by considering an *unknown mode* that does not impose any constraint on the system's variables. The underlying idea is summarized in the following quote of Conan Doyle in the paper of De Kleer and Williams [30] that introduces the Sherlock diagnosis system:

> Sherlock Holmes - The Sign of the Four: "When you have eliminated the impossible, whatever remains, however improbable, must be the truth".

In our context, this would mean that the unknown mode, whatever priori improbable, becomes more probable than the others, since its *no-prediction* is more compatible with observations than the more precise prediction from other modes. We first demonstrated this principle for our hybrid estimation scheme in [54], where we introduced the hybrid system's pendant of the unknown mode, together with a decomposition scheme that offers the capability to detect unforeseen situations in complex multi-component systems and that enables hybrid estimation to continue in a degraded, but fail-safe, manner.

The benefits of the unknown mode capability are mostly relevant for multi-component systems. Nevertheless, we will demonstrate the underlying principles in terms of the flow regulator. We extend the model of the flow regulator with an additional unknown mode that does not specify any difference or algebraic equation among the continuous variables u_c, x_c, and y_c. We integrate this mode in terms of additional transitions from nominal modes to the unknown mode. Figure 2.13 shows the extended transition graph that incor-

porates transitions that traverse the system to the mode unknown[3]. At each
default transition of the nominal modes, we incorporate a transition thread
with low likelihood to the unknown mode. These transition threads model the
fact that the system can fail unexpectedly, regardless of the current mode of
operation[4]. The mode unknown does not specify any algebraic or difference
equation that constrain the continuous variables, thus, we cannot perform
any continuous filtering/estimation operation. Nevertheless, we can incorpo-

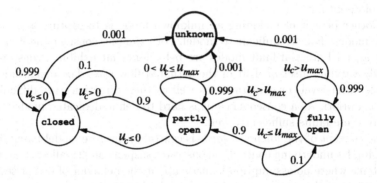

Fig. 2.13. Probabilistic transition graph of flow regulator with unknown mode.

rate this mode into our hybrid estimation scheme by taking the assumption
that the omitted continuous estimation would lead to a perfect match, which
implies the value of $P_O = 1.0$ for the observation function. This specification
allows us to perform the search within the hypothesis tree as usual and enables us to directly compare an unknown mode hypothesis with hypotheses at
nominal modes. The estimation algorithm favors the unknown mode hypothesis over any hypothesis at a nominal mode whenever the product $P_T P_O$ of
the most likely nominal hypothesis is smaller than the transition probability
$P_{T\,\text{unknown}}$ of the unlikely transition to the mode unknown:

$$\max_i P_{T i} P_{O i} < P_{T\,\text{unknown}} \; .$$

Based on the transition model for the flow regulator (Fig. 2.13) this would
mean that hybrid estimation prefers the unknown mode hypothesis whenever

$$\max_i P_{T i} P_{O i} < 0.001 \; .$$

---

[3] For simplicity, we show only transitions *to* the unknown mode but omit the reverse transitions that allow the system to recover from an unknown operational
condition.

[4] Of course, depending on the operational mode we can vary the likelihoods of
an unanticipated failure by varying the probabilities of the associated transition
threads.

Since we are dealing with an operational condition that does not define any constraint among the continuous variables, we cannot provide a grounded estimate for the continuous state. Nevertheless, we want to capture cases, where the system recovers from an unknown operational condition. Therefore, we do maintain a continuous estimate $\hat{x}_c$ for the unknown mode hypothesis with a mean value at its last known value and spread the associated probability density function (PDF) to reflect the continuously decreasing confidence in the continuous estimate (for example, in a Gaussian framework, we hold the mean at he last known estimate $\hat{x}_c$ and continuously increase its variance σ^2 while at mode **unknown**). This allows us to restart the estimation whenever sthe the system recovers from the unknown operational condition.

2.2.5 Focused Real-Time Estimation of Complex Systems

The real advantage of our search-based estimation scheme over traditional multiple-model estimation becomes evident whenever one has to monitor complex real-world artifacts, such as production plants, space probes or modern automotive systems. Those systems are composed of many interconnected components and operate through system-wide complex interaction among those components.

This complex interaction, as well as the overwhelming number of possible failure and operational modes makes it difficult, if not impossible, to apply the traditional multiple-model estimation techniques. Our focused estimation scheme that builds on careful exploration of the large set of possible estimation hypotheses, however, scales much better to systems of high complexity.

In order to efficiently extend the focused search technique that was introduced above to multi-component systems, we have to take the assumption that mode transitions of the system's components at a specific time point are mutually independent. This realistic assumption allows us to view mode transitions of the components individually, so that we can provide an efficient enumeration scheme for possible mode transitions that directly builds upon an extension of the A∗ search based estimation technique. Figure 2.14 illustrates this fact for a multi-component system with ζ components. Instead of having one highly spreading transition expansion and the consecutive filtering operation per estimation step, we do have ζ transition expansions, followed by a single filtering operation for the overall system.

The unknown mode capability is highly valuable for the hybrid estimation of multi-component systems. Individual components at the unknown mode should not prevent us from estimating un-effected parts of the system that are still fully determined in terms of known sensor and actuator signals. For this purpose, we apply a decomposition scheme that clusters the components into subsystems that can be evaluated independently. This limits the level of degradation of the hybrid estimation whenever parts of the system operate at an unknown mode. Estimation can still be done for the components in the fully specified clusters of the system. Our decomposition scheme builds

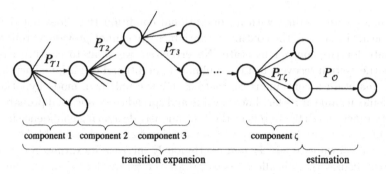

component 1 component 2 component 3   component ζ

transition expansion   estimation

Fig. 2.14. Best-first search for a multi-component system.

upon the principle of causal analysis [83, 96] and structural analysis [41] of
the multi-component system for the mode of a specific estimation hypothesis.

Decomposition contributes to the real-time execution of our hybrid estimation algorithm as well. The large number of modes of the system prevents
us from pre-designing all filters that are required in the course of hybrid estimation. Therefore, our hybrid estimation system deduces extended Kalman
filter on-line and, for efficiency, caches the most recent ones for re-use. The
advantages of performing this operation for a decomposed system are twofold:
Firstly, the cached filters of the component clusters can be used as "building
blocks" for the clustered overall filter. This strategy reduces the number of
filter deductions significantly and increases the utilization of the cached filters.
Secondly, the execution of individual filters for component clusters and the
combination of their estimation results is computationally more efficient than
executing one large filter for the overall system. This is due to the fact that
the computational requirements for a Kalman filter that estimates n state
variables are approximately proportional to n^3 so that several filters for few
state variables outperform a single filter for the overall system.

2.3 Hybrid Estimation in Automation

Process monitoring and diagnosis are essential ingredients of modern automation and control systems. Modern control schemes orchestrate the components of a physical artifact in a complex interleaved way to achieve a desired
high-performance operation. The resulting system-wide interaction makes it
difficult for a human operator to trace and interpret the operation correctly.
Moreover, modern automation and control increasingly handles potentially
dangerous physical artifacts such as chemical plants, power plants, airplanes
or modern automobiles. The loss of control over these artifacts can lead to
severe situations with the potential to harm the environment or endanger human lives. It is, therefore, essential that a monitoring and diagnosis system is
capable of tracing the operation of an artifact in terms of its physical entities

and modes of operation and to robustly cope with atypical situations, such as un-anticipated faults.

Our proposed scheme for hybrid estimation provides some important milestones for such a monitoring and diagnosis system: firstly, a capable component-based modeling paradigm, secondly, a focused estimation algorithm that copes with the complexity of the resulting estimation task and that can operate on-line within a real-time system, and thirdly, explicit incorporation of unknown operational modes that enables us to continue estimation at a possibly degraded, but fail-safe manner.

2.4 Related Research

2.4.1 Stochastic Systems and Kalman Filtering

Real world systems are almost always subject to disturbances, noise, and non-deterministic changes. Therefore, any successful modeling paradigm should not neglect these effects. Stochastic systems theory [1, 84] takes this into account and provides dynamic models, such as sets of ordinary differential or difference equations that are subject to random disturbances. These stochastic models can be used to anticipate the evolution of physical entities of an artifact under investigation from noisy measurements. The Kalman filter [61, 42, 10, 92] for linear stochastic models and the extended Kalman filter [76, 42] for nonlinear stochastic models are the two most prominent members of estimation/filtering algorithms for stochastic systems. An interesting new extension of the Kalman filter to nonlinear systems is the unscented Kalman filter [60].

2.4.2 Hybrid Systems

Hybrid systems research is devoted to modeling, design and validation of systems which can exhibit continuous and discrete modes of behavior. A computer controlled system, where a physical system (plant) is controlled by a

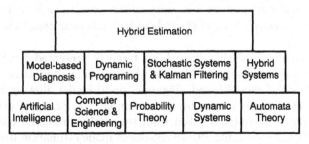

Fig. 2.15. Foundational concepts for hybrid estimation.

computer program involving discrete-time control laws and discrete-event control, is an example for a hybrid system. Hybrid Systems Theory attempts to offer a logical, mathematical, and computational framework for understanding and designing complex heterogeneous systems based on complimentary methodologies that were developed in the fields of Control Theory and Computer Science. As these research communities speak largely different languages and employ largely different methods, hybrid systems research led to a variety of frameworks for modeling, analysis and synthesis of hybrid systems. The paper collections of the *Hybrid Systems Computation and Control (HSCC)* workshop series ([8, 12, 52, 11, 98, 71, 33, 93, 73, 9]) provide a good overview of the developments in this burgeoning research area.

There are many approaches to the task of hybrid systems modeling that mostly depend on the type of analysis or synthesis that one ought to pursue, for example, simulation, controller design, stability analysis, or safety analysis. The approaches differ with respect to the emphasis on the level of complexity that they put onto the continuous and discrete dynamics of the model. Control oriented approaches, for instance, put their emphasis onto complex continuous dynamics (for example, hybrid models for supervisory control [63] or the unified framework for hybrid control of Branicky [23, 22]). Computer science oriented paradigms, on the other hand, emphasize the discrete dynamics, that is, the automaton part of the hybrid model, and utilize simpler continuous dynamics (for example, Alur's timed automata [5]) or emphasize automaton issues such as composition, receptiveness and liveness etc. (for example, see [81]). Most current work, however, can be seen as spanning the spectrum and combines aspects of linear and nonlinear dynamic systems, supervisory control, discrete-event systems, finite automata and Petri nets.

Most hybrid modeling frameworks utilize deterministic models for the continuous dynamics of a system. Only few paradigms, such as [57, 25, 44], deal with stochastic models, as we do in our framework, to capture the disturbances and the noise of real-world systems. Similarly, mode transitions are often modeled either purely deterministically, or non-deterministically but without a probabilistic quantification. Our hybrid model remedies this situation by merging a *hidden Markov Model (HMM)* [18, 36, 86] with a *stochastic discrete-time model* for the continuously valued state variables of the system. In terms of composition of an overall hybrid model for a multi-component system, our modeling framework is in spirit of the reactive module framework of Alur and Henzinger [7, 51] and the hybrid I/O automata framework of Lynch [72, 81].

Several other approaches have been recently presented for hybrid systems estimation and diagnosis. Many of them provide alternative bridges between the fields of model-based diagnosis, multi-model filtering and hybrid systems. Most of these approaches (for example, [78, 82, 110]) utilize a heterogeneous set of techniques taken from the distinct research communities. Reference [110], for example, utilizes Bayesian mode estimation together with a timed Petri net model for diagnosis. Reference [82] combines hybrid behavior tracking, mode estimation and qualitative-quantitative reasoning techniques for fault diagnosis of continuous systems with supervisory control. The hybrid framework for simulation, diagnosis and system tracking that is described in [17] builds upon the same foundation as our work, the Livingstone system [104]. It utilizes interval descriptions to express uncertainties at the continuous level, probabilities for discrete mode estimation and a continuous/discrete interface to synchronize both types of estimation. In contrast, our work provides a homogeneous framework that makes extensive use of probabilities for both, the continuous state and the discrete mode estimation. Other estimation methods that should be noted here build onto less traditional methods such as Bayesian networks [65] or particle filters [77, 66, 31, 64, 32, 40, 99].

2.4.3 Multiple-Model Filtering

Applications in aeronautics and aerospace, in particular target tracking, dealt with the very similar estimation problem of systems that can be in one out of several modes of operation. This lead to a set of estimation algorithms that can be classified as *multiple-model estimation* schemes. Estimation is done by concurrently executing a set of filters (typically, one for each model) and combining the model-conditioned estimates, according to an algorithm-specific weighting scheme. This strategy represents a sub-optimal solution of the hybrid estimation problem as they avoid tracking a set of trajectory hypothesis that is (worst case) exponential in the number of time-steps. Examples of approximate multiple-model estimation schemes are the generalized pseudo-Bayesian (GPB) [2], the detection-estimation [97], the residual correlation Kalman filter bank [48], and the interacting multiple-model algorithm (IMM) [20, 13]. These methods track multiple hybrid estimates over time, but require at least l filters to perform this task, where l is the number of possible models (modes) for the system under investigation.

It was shown in [68] that maintaining an exhaustive set of modes does not only impose an enormous computational burden, but also decreases the prediction quality of the filter as too many, highly unlikely, modes are considered. Adaptive multiple-model estimation was proposed as a possible solution for this dilemma. This estimation scheme adapts the mode-set to a subset of modes that are most likely at a given time-point. Most adaptive multiple-model estimation methods [68, 69, 67] differ in the way in which they obtain an appropriate mode-set. An example for a typical mode-set adaption scheme is to use the transition graph to obtain the set of modes that are immediately reachable, given the previous estimate.

In many complex multi-component systems, however, it is the case that the number possible modes, at a particular time-step, is unrealistically large for an on-line estimation that utilizes standard multiple-model estimation algorithms like IMM or GPBn. Therefore, we apply a different sub-optimal estimation scheme that explores this large number of modes carefully by focusing on the set of most likely modes only.

2.4.4 Dynamic Programming and Best-First Search

One of our contributions is a careful reformulation of the hybrid estimation task in the form of a shortest-path problem. This enables us to apply a wide spectrum of optimization algorithms that are tailored to solving this particular optimization problem. Probably the most prominent algorithm for a shortest-path problem is *Dynamic Programming* [16, 19]. However, the specific structure of our search problem (hybrid estimation leads to an overwhelmingly large number of possible estimation hypotheses and Dynamic Programming would consider all of them) suggests to utilize other search methods that focus onto the set of most likely solutions. The class of *best-first* search methods, out of the toolbox of modern Artificial Intelligence provides a set of algorithms that are highly suitable for our task. In particular, we utilize the A* search [49, 50] and the *beam* search [108, 89] algorithms to solve the reformulated hybrid estimation problem.

2.4.5 Qualitative Reasoning and Model-Based Diagnosis

Qualitative Reasoning [21, 100, 37] has been a successful research branch of Artificial Intelligence since the early 1980's. Qualitative reasoning methods intend to replicate, in the computer, parts of human reasoning for the tasks of modeling, simulation, and causal explanation of uncertain dynamic systems, that is, systems where it is difficult to provide precise and complete mathematical models (the paper collections cited above provide material for each of these subjects). The ability of qualitative reasoning methods to predict and explain the behavior of uncertain dynamic systems makes them a natural choice for monitoring and diagnosis [29, 79]. Research in this particular direction is known as *Model-based Diagnosis* [46]. Its underlying concept of constraint suspension [28] allows diagnosis of systems where no assumption is made about the behavior of one or several components of the system. We utilize this concept, together with efficient methods for causal analysis [83, 96], to incorporate the principle of the *unknown mode* in our hybrid estimation scheme. This significantly extends our basic hybrid estimation algorithm that has its roots in the *Livingstone* [104] model-based diagnosis and reactive control system.

Interesting applications of discrete model-based diagnosis are, for instance, monitoring and diagnosis of gas turbines [95], automotive systems [90], industrial applications [85, 88], and a space-probe (Livingstone) [105].

2.4.6 Fault Detection and Isolation (FDI)

A traditional approach to fault detection is based on hardware redundancy methods, which use multiple lanes of sensors and actuators with a voting scheme to decide, whether and when a fault has occurred. In contrast to this cost intensive approach used in safety-critical applications, it is possible to utilize information provided by single sensors and the mathematical model of the underlying system which provides functional relationships between the measured variables of the monitored system. Such a redundancy concept that relies on information of dissimilar measured variables and on the mathematical model of the system is called *analytical (functional) redundancy* and a fault detection and isolation (FDI) scheme based on this methodology is called a *model-based FDI* scheme. The methods utilize the analytic redundancy in that they compare measurements with an estimation that is based on the mathematical model. The resulting difference specifies a *residual* signal that is zero if the checked component of the system is operating normally, and non-zero in the event of a fault. Fault detection and isolation utilizes the residuals, detects fault situations whenever they exceed a certain threshold, and isolates the fault using some sort of decision logic.

The idea of model-based FDI, that is, replacing hardware redundancy by analytical redundancy, goes back to Beard [15], who formulated this approach at the MIT in the early 1970s. Refinements done by Jones led to the *Beard-Jones Fault Detection Filter* [59]. In parallel with this development, methods based on statistical approaches were developed in the early 1970s. These model-based FDI concepts with their emphasis on stochastic systems and jump detection are summarized in the survey papers [107, 14]. Clark et al. first applied Luenberger observers for fault detection [27] and developed various sensor fault isolation schemes. Developments in observer-based methods for model-based FDI are summarized in a survey paper [38]. Another important FDI approach proposed in the late 1970s is to apply parameter estimation methods. Research results of this FDI approach are described in a comprehensive survey paper [58]. Methods based on consistency checking of system input and output data over a time window are reported as parity equation approaches in literature [43].

All model-based FDI schemes described rely on a mathematical model of the monitored dynamic system. However, mathematical models will almost always represent approximations of the real world so that an exact agreement between the model and the monitored dynamic system cannot be achieved. This modeling uncertainty, together with the disturbances and noise experienced in real-world applications, emphasizes the necessity for *robust* model-based FDI methodologies which can cope with modeling uncertainty and disturbances/noise. This issue received much attention in recent FDI research and led to robust extensions of the methodologies given above (for example, see [26] for a recent monograph on this topic).

3

Probabilistic Hybrid Automata

The previous chapter introduced our hybrid modeling formalism informally. We shall now proceed with a detailed definition of the underlying hybrid model that builds the basis for the remaining parts of this monograph. Our aim is to model artifacts that are composed of many individual components, each of them with several modes of operation and failure. The overall behavior of the artifacts is a result of complex interaction among those components and shows a mixture of continuous evolution and discrete changes. This leads naturally to a *component-based* modeling paradigm, where individual components are modeled as *hybrid automata*. *More precisely, probabilistic hybrid automata*, since our aim is to estimate the complex behavior of an artifact that is subject to external disturbances (noise), unforeseen failures, and un-anticipated environmental interactions.

To model the individual components of an artifact, we start by using a *Hidden Markov Model (HMM)* [18, 36, 86] to describe discrete stochastic changes in the system. We will then fold in the continuous dynamics, by associating a set of dynamic and algebraic equations with each HMM mode[1]. This will lead to a model that we call *probabilistic hybrid automaton, or PHA* for short [102, 55]. Component models will be then strung together according to the artifacts blueprints and lead to the *concurrent probabilistic hybrid automaton (cPHA)* that describes the overall system in terms of the component models, their composition and the interconnection to the outside world.

3.1 Hidden Markov Models

Let us start with a simple generic component with three possible modes {standby, on, fault}. The mode is *hidden* and cannot be observed directly.

---

[1] To avoid confusion in terminology, we refer to the HMM state as the *mode* and reserve the term state to refer to the state of the overall probabilistic hybrid automaton.

An observation of the component only reveals whether the component works (modes standby and on) or whether it experiences a fault. These two observations are indicated in terms of the discrete values {ok, faulty}. The behavior of the component can be modeled as a stochastic automaton model with mode variable x_d and observation variable y_d. The model captures observations and mode transitions probabilistically in terms of the probabilistic transition function P_T and the probabilistic observation function P_O that specify the conditional probabilities

$$P_T(x_{d,k}, x_{d,k-1}) := P(x_{d,k}|x_{d,k-1})$$
$$P_O(y_{d,k}, x_{d,k}) := P(y_{d,k}|x_{d,k}) \,, \tag{3.1}$$

where $x_{d,k} \in \{\text{standby, on, fault}\}$ and $y_{d,k} \in \{\text{ok, faulty}\}$ denote the valuations of x_d and y_d at time-step k, respectively. The mode transitions satisfy the *Markov property* as P_T describes a probability that is only conditioned on the previous mode $x_{d,k-1}$. This fact, together with the *hiding* property of the observation process, justifies the name *Hidden Markov Model* for the stochastic automaton.

Figure 3.1a visualizes the modes (nodes) and transitions (arcs) for the HMM in graphical notation. The possible transitions are labeled with their associated transition probability. For instance, whenever the HMM is at the mode on, it remains at this mode with probability 0.8, or transitions to mode standby with probability 0.15, or to mode fault with probability 0.05.

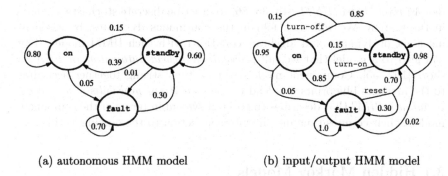

(a) autonomous HMM model   (b) input/output HMM model

Fig. 3.1. Visualization of Hidden Markov Models (HMM) as graphs.

Up to now, we view transitions as purely autonomous. They only depend on the mode of the stochastic automaton. This *autonomous HMM* is the usual form of the Hidden Markov Model as it can be found in literature (e.g. [36, 86]). Many applications, however, require additional inputs to capture commanded transitions. For instance, in our example we could think of a command input variable u_d that can take on a discrete value from the set {turn-on, turn-off, reset, no-command} to command mode changes[2]. Since we intend to model a real-world system it is desirable to specify the outcome of a command probabilistically. For example, whenever the system is in mode standby and experiences the turn-on command it will change to the mode on with high probability, whereas there is a small chance that the command shows no effect and the system remains at the mode standby. This situation is visualized in Fig. 3.1b, where a command $u_{d,k} =$ turn-on *guards* the transition that leads to the mode $x_{d,k+1} =$ on with probability 0.85 and to the mode standby with probability 0.15. The unlabeled transition captures all other cases $u_{d,k} \in \{$turn-off, reset, no-command$\}$. It specifies that the system remains in the mode standby with probability 0.98, or transitions to the mode fault with probability 0.02.

HMMs with inputs are sometimes called *input/output Hidden Markov Models (IOHMM)* in literature [18]. We do not want to make this distinction here and will refer to both types of stochastic automata as HMMs. The following definition of an HMM will be the basis for our stochastic hybrid model that we develop and use throughout this thesis.

Definition 3.1. A *Hidden Markov Model (HMM)* can be described by a tuple

$$\langle x_d, u_d, y_d, \mathcal{X}_d, \mathcal{U}_d, \mathcal{Y}_d, P_\Theta, P_\mathcal{T}, P_\mathcal{O} \rangle .$$

The variables x_d, u_d and y_d with discrete domains $\mathcal{X}_d = \{m_1, \ldots, m_l\}$, $\mathcal{U}_d = \{u_1, \ldots, u_\nu\}$ and $\mathcal{Y}_d = \{y_1, \ldots, y_\psi\}$ denote the *mode, command input* and *observation*, respectively. Their valuations at time-step k shall be denoted by $x_{d,k}, u_{d,k}$ and $y_{d,k}$. The *initial mode function* P_Θ specifies a probability mass function $p_d(x_{d,0})$ among the modes $m_i \in \mathcal{X}_d$. The *mode transition function* $P_\mathcal{T}$ describes the conditional probability $P(m_i|u_\varsigma, m_j)$ of transitioning from mode $x_{d,k-1} = m_j$ to $x_{d,k} = m_i$, given the discrete command input $u_{d,k-1} = u_\varsigma$. The *observation function* $P_\mathcal{O}$ describes the conditional probability $P(y_i|u_\varsigma, m_j)$ that one observes the discrete value $y_{d,k} = y_i$, given the discrete command input $u_{d,k} = u_\varsigma$ and the mode $x_{d,k} = m_j$.

The definition (3.1) specifies mode transitions in terms of a multivariate conditional probability mass function that characterizes $P_\mathcal{T}$. One could interpret the transitions as being *guarded* by propositional formulas such as $u_d =$ turn-on for the transition standby $\rightarrow \{$standby, on$\}$ or $(u_d =$ turn-off$) \vee (u_d =$ reset$) \vee (u_d =$ no-command$)$ for the unlabeled transition standby $\rightarrow \{$standby, fault$\}$. This leads to an alternative transition description in terms of a set valued function $T(x_d)$. This function provides for each mode $m_i \in \mathcal{X}_d$ a set of *transition tuples* $\langle p_{\tau i}, c_i \rangle =: \tau_i$. c_i denotes the guard and the probability mass function $p_{\tau i}$ specifies the probabilities of the transition threads, given that the guard is satisfied. Figure 3.2 visual-

---

[2] For symmetry, we treat the non-transition $x_{d,k} = m_j \rightarrow x_{d,k+1} = m_j$ as a transition.

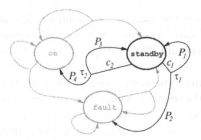

Fig. 3.2. Guarded probabilistic mode transitions.

izes this concept for the mode **standby**. $T(\mathbf{standby})$ specifies the transition set $\{\langle p_{\tau 1}, c_1 \rangle, \langle p_{\tau 2}, c_2 \rangle\}$. The transition $\tau_2 = \langle p_{\tau 2}, c_2 \rangle$ denotes a transition from the mode **standby** to itself with probability $P_3 = p_{\tau 2}(\mathbf{standby})$, or to the mode **on** with probability $P_4 = p_{\tau 2}(\mathbf{on})$, whenever its associated guard $c_2 := (u_d = \mathbf{turn\text{-}on})$ is satisfied.

In the following specification of our stochastic hybrid automaton model we will favor the latter transition definition to encode a probabilistic transition function P_T.

3.2 Probabilistic Hybrid Automata

The Hidden Markov Model abstracts physical entities into discretely valued variables. However, we intend to capture the behavior of a physical artifact in terms of discretely valued variables for the modes of operation and continuously valued variables that capture the continuous evolution of physical entities. Therefore, we use an HMM model as the automaton basis for our hybrid model and supplement it with additional continuous variables and a mode-dependent specification of the continuously valued dynamics.

The major application of our stochastic automaton model is mode and state estimation, in the context of process monitoring and diagnosis. This task will almost exclusively execute on a supervisory control system in discrete-time. Therefore, we restrict the class of dynamic models to *discrete-time models* and describe the dynamic evolution of the continuously valued state and I/O variables in terms of *difference equations* and *algebraic equations*. Of course, this implies a certain level of abstraction, where interconnected continuous time processes of the artifact are individually abstracted in terms of discrete-time equations that capture sufficient detail for the monitoring and diagnosis task.

In general, this abstraction could be relaxed and it is possible to utilize both, continuous time and discrete-time models (possibly with different sampling-times) for the individual components within one hybrid model. However, this would involve a hybrid estimation framework that is unnecessarily detailed and complex. As a consequence, we decided to restrict the following presentation to discrete-time systems with one common sampling rate.

As a preliminary step toward defining our stochastic hybrid model for a complex artifact, we first define the *probabilistic hybrid automaton (PHA)* model for individual system components. PHA models of the system components will serve as the building blocks for our overall modeling paradigm, the *concurrent probabilistic hybrid automata (cPHA)*. More specifically, we define the component automaton model as:

Definition 3.2. A *discrete-time probabilistic hybrid automaton (PHA)* $\mathcal{A}$ can be described as a tuple $\langle \mathbf{x}, \mathbf{w}, F, T, X_0, \mathcal{X}_d, \mathcal{U}_d, T_s \rangle$:

- $\mathbf{x}$ denotes the hybrid *state variables* of the automaton[3], composed of $\mathbf{x} = \mathbf{x}_d \cup \mathbf{x}_c$. The discretely valued state variables $\mathbf{x}_d = \{x_{d1}, \ldots, x_{dn_m}\}$ denote the *mode* of the automaton and have finite domain $\mathcal{X}_d = \{\mathbf{m}_1, \ldots, \mathbf{m}_l\}$. The continuously valued state variables $\mathbf{x}_c = \{x_{c1}, \ldots, x_{cn_x}\}$ capture the dynamic evolution of the automaton with domain $\mathbb{R}^{n_x}$.

- The set of *I/O variables* $\mathbf{w} = \mathbf{u}_d \cup \mathbf{u}_c \cup \mathbf{v}_c \cup \mathbf{y}_c$ of the automaton is composed of disjoint sets of discrete input variables $\mathbf{u}_d = \{u_{d1}, \ldots, u_{dn_c}\}$ (also called *command variables*), continuous *input variables* $\mathbf{u}_c = \{u_{c1}, \ldots, u_{cn_u}\}$, continuous *disturbances* $\mathbf{v}_c = \{v_{c1}, \ldots, v_{cn_v}\}$ and continuous *output variables* $\mathbf{y}_c = \{y_{c1}, \ldots, y_{cn_y}\}$. We use the notation $\mathbf{w}_c := \mathbf{u}_c \cup \mathbf{v}_c \cup \mathbf{y}_c$ to denote the set of continuously valued I/O variables. The I/O variables have the domains $\mathcal{U}_d$, $\mathbb{R}^{n_u}$, $\mathbb{R}^{n_v}$, and $\mathbb{R}^{n_y}$, respectively.

- The set-valued function[4] $F : \mathcal{X}_d \rightarrow 2^{\mathcal{F}_{DE}} \times 2^{\mathcal{F}_{AE}}$ specifies the *continuous evolution* of the automaton in terms of sets of *discrete-time difference equations* $F_{DE} \subseteq \mathcal{F}_{DE}$ and *algebraic equations* $F_{AE} \subseteq \mathcal{F}_{AE}$ for each mode $\mathbf{m}_j \in \mathcal{X}_d$. T_s denotes the sampling-period of the discrete-time difference equations.

- The set-valued function $T : \mathcal{X}_d \rightarrow 2^{\mathcal{T}}$ specifies the probabilistic discrete evolution of the automaton in terms of a finite set of *transition triples* $\tau_i := \langle p_{\tau i}, c_i, r_i \rangle \in \mathcal{T}$. Each *transition specification* τ_i has an associated Boolean *guard* $c_i : \mathbb{R}^{n_x} \times \mathbb{R}^{n_u} \times \mathcal{U}_d \rightarrow \{\text{true}, \text{false}\}$ and specifies the probability mass function $p_{\tau i} : \mathcal{X}_d \rightarrow [0\ 1]$ over *transition target modes* $\mathbf{m}_j \in \mathcal{X}_d$. The third component, r_i, handles the valuation of the continuous states across the transition. A transition can either *inherit* the continuous state or cause a *reset* to a new value. The function $r_i : \mathcal{X}_d \times pdf \rightarrow pdf$ specifies inheritance or a change of the continuous state valuation from $\mathbf{x}_{c,k}$ to a new value $\mathbf{x}'_{c,k}$, immediately after the transition.

- The initial state of the automaton is specified in terms of the tuple $X_0 := \langle p_{d0}, P_{c0} \rangle$. The function $p_{d0} : \mathcal{X}_d \rightarrow [0\ 1]$ specifies the initial probability

---

[3] When clear from context, we use lowercase bold symbols, such as $\mathbf{v}$, to denote a *set* of variables $\{v_1, \ldots, v_l\}$, as well as a *vector* $[v_1, \ldots, v_l]^T$ with components v_i.

[4] We use the notation $2^{\mathcal{S}}$ to denote the *power set* of a set $\mathcal{S}$, that is, the collection of *all subsets*: $2^{\mathcal{S}} = \{X | X \subseteq \mathcal{S}\}$.

mass among the modes, and $P_{c0} : \mathcal{X}_d \rightarrow pdf$ specifies for every mode $\mathbf{m}_j \in \mathcal{X}_d$ a multivariate probability distribution function p_{c0} for the continuous state $\mathbf{x}_{c,0}$.

Table 3.1 summarizes the naming conventions for the I/O variables of a PHA and visualizes their separation into disjoint sets.

Table 3.1. Classification of the I/O variables.

w		
$\mathbf{w}_d$	$\mathbf{w}_c$	
u	v	y
u_d u_c	v_c	y_c
independent I/O variables	dependent I/O variables	

The transition specifications τ_i express probabilistic transitions that are guarded on the continuous state $\mathbf{x}_c$, the discrete input $\mathbf{u}_d$, and possibly the continuous input[5] $\mathbf{u}_c$. The guards $c_i(\mathbf{x}_c, \mathbf{u}_c, \mathbf{u}_d)$ are of the form

$$\left(b_x^- \le q_{xc}(\mathbf{x}_c) < b_x^+\right) \wedge \left(b_u^- \le q_{uc}(\mathbf{u}_c) < b_u^+\right) \wedge c_{ud}(\mathbf{u}_d) , \qquad (3.2)$$

where $c_{ud}(\cdot)$ is a propositional logic formula and $q_{xc}(\cdot)$, $q_{uc}(\cdot)$ specify nonlinear functions. The bounds b_x^-, b_x^+, b_u^-, and b_u^+ denote constant boundary values that can take on values of the *extended* real number line, $\mathbb{R}^*$, which includes the endpoints $-\infty$ and ∞, for example:

$$(-\infty \le 1 + x_{c1} + x_{c1}^3 - x_{c2} < 0) \wedge (0 \le u_{c1} < 1) \wedge (u_{d1} = \textbf{turn-on}) . \quad (3.3)$$

The specification of the continuous evolution in terms of the function $F(\cdot)$ that specifies sets of difference and algebraic equations will be particularly useful in the context of multi-component automata (cPHA). The system-wide context will specify which variable $w_c \in \mathbf{w}_c$ serves as input and which one serves as output to the component automaton. Nevertheless, let us provide more details on the component F of the PHA here as well. Consider a PHA

$$\mathcal{A} = \langle \{x_{d1}, x_{c1}, x_{c2}\}, \{u_{d1}, w_{c1}, w_{c2}, v_{c1}, v_{c2}\}, F, T, \{m_1, m_2, m_3\}... \rangle , \quad (3.4)$$

with the following set of equations for the mode m_1:

$$F_{DE}(m_1) = \{x_{c1,k+1} = 0.6 \, x_{c1,k} + w_{c1,k} + v_{c1,k}$$
$$x_{c2,k+1} = 0.1 \, x_{c1,k} + 0.8 \, x_{c2,k} + v_{c2,k}\}, \qquad (3.5)$$
$$F_{AE}(m_1) = \{w_{c2} = 1.0 \, x_{c2} + w_{c1,k}\} . \qquad (3.6)$$

---

[5] Our PHA specification does not specify a priori, whether a variable $w_{ci} \in \mathbf{w}_c$ acts as an input or output. Using an I/O variable w_{ci} for a transition guard, however, implies that this variable acts as an input, i.e. $w_c \in \mathbf{u}_c$, for the PHA at the particular mode for which the transition is specified.

If we assume that $w_{c1,k}$ is an input variable and $w_{c2,k}$ is an output variable ($\mathbf{u}_c = [w_{c1}]$ and $\mathbf{y}_c = [w_{c2}]$) we obtain the following mathematical model for $\mathcal{A}$:

$$\mathbf{x}_{c,k+1} = \begin{bmatrix} 0.6 & 0 \\ 0.1 & 0.8 \end{bmatrix} \mathbf{x}_{c,k} + \begin{bmatrix} 1 \\ 0 \end{bmatrix} \mathbf{u}_{c,k} + \mathbf{v}_{c,k} \tag{3.7}$$

$$\mathbf{y}_{c,k} = \begin{bmatrix} 0 & 1 \end{bmatrix} \mathbf{x}_{c,k} + \mathbf{u}_{c,k} . \tag{3.8}$$

The vector $\mathbf{x}_c = [x_{c1}, x_{c2}]^T$ denotes the continuous state and the vector $\mathbf{v}_c = [v_{c1}, v_{c2}]^T$ denotes continuous disturbances that act upon the state variables of the PHA.

The PHA definition does not predefine the separation of the continuous I/O variables $\mathbf{w}_c$ into the disjoint sets of inputs ($\mathbf{u}_c$), outputs ($\mathbf{y}_c$), and disturbances ($\mathbf{v}_c$). The separation for a particular mode is subject to the interconnection of the PHA component with the outside world and the set of equations $F(m_j)$. Both specifications imply a particular, mode dependent, causality among the continuous variables that separates the I/O variables into disjoint sets of independent (input), noise, and dependent (output) variables. This increases the expressiveness of our modeling framework in that we enable the model to capture changes in the causal structure of the overall system. Reconsider the PHA (3.4) at mode m_1. The associated set of equations does not enforce the particular causality where w_{c1} is the independent variable and w_{c2} is the dependent I/O variable. If we connect the PHA component to an environment that determines w_{c2}, we would obtain a mathematical model with $\mathbf{u}_c = [w_{c2}]$ and $\mathbf{y}_c = [w_{c1}]$, more specifically we would obtain an alternative mathematical model:

$$\mathbf{x}_{c,k+1} = \begin{bmatrix} 0.6 & -1 \\ 0.1 & 0.8 \end{bmatrix} \mathbf{x}_{c,k} + \begin{bmatrix} 1 \\ 0 \end{bmatrix} \mathbf{u}_{c,k} + \mathbf{v}_{c,k}$$
$$\mathbf{y}_{c,k} = \begin{bmatrix} 0 & -1 \end{bmatrix} \mathbf{x}_{c,k} + \mathbf{u}_{c,k} . \tag{3.9}$$

The potential to reverse the input/output directionality for the model $F(m_1)$ (3.5)-(3.6) is due to the dependency of w_{c2} on the continuous state variable x_{c2} *and* the I/O variable w_{c1} in the algebraic equation (3.6). This dependency leads to a discrete-time model with a direct transmission term in the output equation (3.8), and as a consequence, implies the non-unique causality of the system. Whenever the set of equations for a particular mode lacks this direct transmission property, we obtain a unique causality, i.e. the equations of the PHA specify which I/O variable serves as an input to the PHA and which one serves as an output. For example, consider the mode m_2 that implies the following set of equations:

$$F_{DE}(m_2) = \{x_{c1,k+1} = 0.4\, x_{c1,k} + w_{c1,k} + v_{c1,k},$$
$$x_{c2,k+1} = 0.2\, x_{c1,k} + 0.5\, x_{c2,k} + v_{c2,k}\}, \tag{3.10}$$
$$F_{AE}(m_2) = \{w_{c2} = 0.5\, x_{c2}\} . \tag{3.11}$$

The algebraic equation (3.11) lacks the direct dependency among the I/O variables w_{c1} and w_{c2}. As a consequence, we obtain the particular causality $w_{c1} \to \ldots \to w_{c2}$. Thus, the equations for the mode m_2 imply that w_{c1} is the input variable and w_{c2} is the output variable of the cPHA.

PHA composition

We now introduce the *composition* operation for probabilistic hybrid automata. This operation enables us to describe a complex system as composite automaton, where individual components are represented by PHA automata. Several formalisms for the composition of hybrid automata were proposed recently, e.g. [4, 6, 7, 51, 72, 81]. Our composition operation is in spirit of the reactive module framework of Alur and Henzinger [7, 51] and establishes the interconnection of component automata via shared *continuous I/O variables* $w_{c\gamma} \in \mathbf{w}_c$. More precisely, a variable $w_{c\gamma}$ connects an automaton $\mathcal{A}_i$ with the automaton $\mathcal{A}_j$ if $w_{c\gamma} \in \mathbf{w}_{ci}$ and $w_{c\gamma} \in \mathbf{w}_{cj}$. The sets $\mathbf{w}_{ci}$ and $\mathbf{w}_{cj}$ denote the I/O variables of automaton $\mathcal{A}_i$ and $\mathcal{A}_j$, respectively. In terms of set operations we can concisely express this property as $w_{c\gamma} \in \mathbf{w}_{ci} \cap \mathbf{w}_{cj}$.

Of course, we can only apply composition to two automata $\mathcal{A}_1$ and $\mathcal{A}_2$ if they are *compatible*. Compatibility in this context involves three properties: Firstly, we do not allow automata to share state variables $x_c \in \mathbf{x}_{ci}$, secondly, we require that the automata have distinct disturbance variables $v_c \in \mathbf{v}_{ci}$, and thirdly, we demand that both automata operate at the same sampling period T_s.

In the following we denote the components of a PHA specification $\mathcal{A}_j$ by using the same subscript, that is, $\mathbf{x}_j = \mathbf{x}_{dj} \cup \mathbf{x}_{cj}$, $\mathbf{w}_j = \mathbf{u}_{dj} \cup \mathbf{u}_{cj} \cup \mathbf{v}_{cj} \cup \mathbf{y}_{cj}$, $F_j, T_j, X_{0j}, \mathcal{X}_{dj}, \mathcal{U}_{dj}, T_{sj}$. With this notation we can define PHA compatibility and PHA composition as follows:

Definition 3.3 (PHA compatibility). Let $\mathbf{z}_{cj} = \mathbf{x}_{cj} \cup \mathbf{u}_{cj} \cup \mathbf{y}_{cj} \cup \mathbf{v}_{cj}$ denote the set of all continuous variables of an automaton $\mathcal{A}_j$. Then, we call two probabilistic hybrid automata $\mathcal{A}_1$ and $\mathcal{A}_2$ *compatible*, if $\mathbf{x}_{ci} \cap \mathbf{z}_{cj} = \mathbf{v}_{ci} \cap \mathbf{z}_{cj} = \emptyset$ for $i \neq j$, and $T_{s1} = T_{s2}$.

Definition 3.4 (PHA composition). The (concurrent) *composition* $\mathcal{A}_1 \parallel \mathcal{A}_2$ of two compatible probabilistic hybrid automata $\mathcal{A}_1$ and $\mathcal{A}_2$ is defined in terms of the tuple

$$\langle \mathbf{x}, \mathbf{w}, F, T, X_0, \mathcal{X}_d, \mathcal{U}_d, T_s \rangle,$$

where: $\mathbf{x} := \mathbf{x}_d \cup \mathbf{x}_c$, with $\mathbf{x}_d := \mathbf{x}_{d1} \cup \mathbf{x}_{d2}$, $\mathbf{x}_c := \mathbf{x}_{c1} \cup \mathbf{x}_{c2}$,

$\mathbf{w} := \mathbf{w}_1 \cup \mathbf{w}_2$,

F : $F(\mathbf{x}_d) := F_1(\mathbf{x}_{d1}) \cup F_2(\mathbf{x}_{d2})$,

T : $T(\mathbf{x}_d) := T_1(\mathbf{x}_{d1}) \times T_2(\mathbf{x}_{d2})$,

$X_0 := \langle p_{d0}, P_{c0} \rangle$, with $p_{d0}(\mathbf{x}_d) := p_{d01}(\mathbf{x}_{d1}) \cdot p_{d02}(\mathbf{x}_{d2})$,

and $P_{c0}(\mathbf{x}_d) := P_{c01}(\mathbf{x}_{d1}) \times P_{c02}(\mathbf{x}_{d2})$,

$\mathcal{X}_d := \mathcal{X}_{d1} \times \mathcal{X}_{d2}$,

$\mathcal{U}_d := \mathcal{U}_{d1} \times \mathcal{U}_{d2}$,

$T_s := T_{s1}$.

The composition operation ensures that the composition $\mathcal{A}_1 \parallel \mathcal{A}_2$ of two PHAs, again is a PHA $\mathcal{A}$. To emphasize that an automaton $\mathcal{A}$ denotes the composition $\mathcal{A}_1 \parallel \mathcal{A}_2$, we will also call $\mathcal{A}$ a *composite* automaton. A consequence from this property of composition is that we can obtain the composition $\mathcal{A} = \mathcal{A}_1 \parallel \mathcal{A}_2 \parallel \cdots \parallel \mathcal{A}_\zeta$ of ζ compatible automata recursively:

$$\mathcal{A} = (\ldots((\mathcal{A}_1 \parallel \mathcal{A}_2) \parallel \mathcal{A}_3) \parallel \cdots \parallel \mathcal{A}_\zeta) .$$

Let us demonstrate the composition operation of two PHAs $\mathcal{A}_1$ and $\mathcal{A}_2$ with the following example:

$$\mathcal{A}_1 = \langle \{x_{d1}, x_{c1}\}, \{u_{d1}, w_{c1}, w_{c2}, w_{c3}, v_{c1}\}, F_1, T_1, \{m_{11}, m_{12}, m_{13}\}...\rangle$$
$$\mathcal{A}_2 = \langle \{x_{d2}, x_{c2}, x_{c2}\}, \{u_{d1}, u_{d2}, w_{c2}, w_{c3}, w_{c4}, v_{c2}, v_{c3}, v_{c4}\}, F_2, T_2, \{m_{21}, m_{22}\}...\rangle$$
$$(3.12)$$

The automata are compatible since their continuous state and noise variables are distinct (we assume equality of their sampling rates T_{s1} and T_{s2}). The composition $\mathcal{A}_1 \parallel \mathcal{A}_2$ leads to a composite PHA $\mathcal{A}$ with the mode variables $\mathbf{x}_d = \{x_{d1}, x_{d2}\}$ and the continuous state variables $\mathbf{x}_c = \{\mathbf{x}_{c1}, \mathbf{x}_{c2}, \mathbf{x}_{c3}\}$. The interconnection between the automata is achieved by linking the shared I/O variables u_{d1}, w_{c2}, w_{c3}, as shown in Fig. 3.3 . Note that the composition does not imply a particular directionality of the interconnections. The directionality is subject to the causality in the system and depends on the equations for a particular mode assignment and the interconnection to the outside world. For example, consider a PHA mode $\mathbf{x}_{dk+1} = [m_{11}, m_{21}]^T$, which indicates that $\mathcal{A}_1$ is at the mode m_{11}, and $\mathcal{A}_2$ is at the mode m_{21}. This mode assignment leads to the set of equations $F(\mathbf{x}_{dk+1}) = F_1(m_{11}) \cup F_2(m_{21})$, where F_1 and F_2 are given as follows:

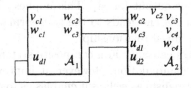

Fig. 3.3. PHA composition example.

$$F_1(m_{11}) = \{x_{c1,k+1} = 0.4\, x_{c1,k} + w_{c1,k} + w_{c2,k} + v_{c1,k},$$
$$w_{c3} = 0.3\, x_{c1}\}$$
$$F_2(m_{21}) = \{x_{c2,k+1} = x_{c2,k} + w_{c2,k} + v_{c2,k}, \qquad\qquad (3.13)$$
$$x_{c3,k+1} = 0.5\, x_{c3,k} + w_{c2,k} + v_{c3,k},$$
$$w_{c3} = 0.2\, x_{c2} + w_{c2}, \quad w_{c4} = x_{c2} + x_{c3} + v_{c4}\} \,.$$

Although we still have to solve the set of equations symbolically to obtain a mathematical model of the usual form

$$x_{c,k+1} = f(x_{c,k}, w_{c,k}, v_{c,k})$$
$$w_{c,k} = g(x_{c,k}, w_{c,k}, v_{c,k}) \,, \qquad\qquad (3.14)$$

we interpret the set of equations (3.13) as the *raw model* for the system at the mode $x_{d\,k+1} = [m_{11}, m_{21}]^T$. Our implementation of the hybrid estimation framework applies a symbolic solver that is based on causal analysis [83, 96] and Gröbner Bases [24], and perform the algebraic manipulations on-line in the course of estimation[6]. The solver transforms the raw model (3.13) into the following set of difference and algebraic equations:

$$x_{c1,k+1} = 0.7\, x_{c1,k} - 0.2\, x_{c2,k} + w_{c1,k} + v_{c1,k}$$
$$x_{c2,k+1} = 0.3\, x_{c1,k} + 0.8\, x_{c2,k} + v_{c2,k}$$
$$x_{c3,k+1} = 0.3\, x_{c1,k} - 0.2\, x_{c2,k} + 0.5\, x_{c3,k} + v_{c3,k}$$
$$w_{c1,k} = \text{exogenous} \qquad\qquad (3.15)$$
$$w_{c2,k} = 0.3\, x_{c1,k} - 0.2\, x_{c2,k}$$
$$w_{c3,k} = 0.3\, x_{c1,k}$$
$$w_{c4,k} = x_{c2,k} + x_{c3,k} + v_{c4,k} \,.$$

The causal analysis identifies the I/O variable w_{c1} as *exogenous* or *independent* variable of the PHA composition $\mathcal{A}_1 \parallel \mathcal{A}_2$ and w_{c2} serves as an output for $\mathcal{A}_2$ and as an input for $\mathcal{A}_1$. This causal specification cannot be seen from the raw equations $F_2(m_{21})$ for the automaton component $\mathcal{A}_2$. Looking at $F_2(m_{21})$ directly, one would expect that w_{c2} serves as an input of $\mathcal{A}_2$ and w_{c3}, w_{c4} are the output variables of $\mathcal{A}_2$. However, the dynamics in the automaton component $\mathcal{A}_1$ enforce that w_{c3} is a dependent variable of $\mathcal{A}_1$. This is due to the fact that w_{c3} only depends on the state variable x_{c1} of $\mathcal{A}_1$. Thus, w_{c3} serves as continuous output for $\mathcal{A}_1$. This implies that w_{c2} and w_{c4} are dependent variables for $\mathcal{A}_2$ and leads to the transformation of the equations $F_2(m_{21})$ to reflect this causal relationship.

The ability to change causality in a composite system requires additional analysis and symbolic manipulation. However, it improves expressiveness of our hybrid modeling formalism as well as it enables us to deal with operational

---

[6] Our current implementation restricts the type of algebraic equations and nonlinear functions to expressions that enable symbolic solutions in explicit form.

conditions that impose changed causality in the system. The latter fact represents a significant advantage over other modeling paradigms for monitoring and diagnosis and justifies the extra computational requirements.

3.3 Concurrent Probabilistic Hybrid Automata

The composition of PHAs $\mathcal{A}_1 \parallel \ldots \parallel \mathcal{A}_\zeta$ specifies the component models and their interconnection via shared variables. It does not specify the interconnection to the outside world. This is the task of the *concurrent Probabilistic Hybrid Automaton (cPHA)* specification. A cPHA takes a composite PHA and specifies the I/O variables that connect the system to the outside world, as well as it quantifies the disturbances in terms of their probability distributions. More precisely:

Definition 3.5. A *concurrent probabilistic hybrid automaton (cPHA)* $\mathcal{CA}$ can be described as a tuple $\langle \mathcal{A}, \mathbf{u}, \mathbf{y}_c, \mathbf{v}_c, N \rangle$:

- $\mathcal{A} = \mathcal{A}_1 \parallel \mathcal{A}_2 \parallel \ldots \parallel \mathcal{A}_\zeta$ denotes the composite PHA, comprised of the PHA models $\mathcal{A}_i$ for the individual components.
- The *input variables* $\mathbf{u} = \mathbf{u}_d \cup \mathbf{u}_c$ of the automaton consists of the set of discrete input variables $\mathbf{u}_d = \mathbf{u}_{d1} \cup \ldots \cup \mathbf{u}_{d\zeta}$ (command variables) and the set of continuous input variables $\mathbf{u}_c \subseteq \mathbf{w}_c = \mathbf{w}_{c1} \cup \ldots \cup \mathbf{w}_{c\zeta}$ that is determined by the outside world. Alternatively, we will also call the continuous input variables $\mathbf{u}_c$ the *independent* variables of $\mathcal{A}$.
- The *output variables* $\mathbf{y}_c \subseteq \mathbf{w}_c$ specify the subset of observed continuous I/O variables of $\mathcal{A}$.
- The *noise variables* $\mathbf{v}_c \subseteq \mathbf{w}_c$ specify the the subset of continuous I/O variables that model the disturbances that act upon the system. The cPHA specification quantifies the disturbances in terms of the mode dependent[7] function $N : \mathcal{X}_d \to pdf$ that specifies a multivariate probability density function p_v for the noise variables $\mathbf{v}_c$.

The Fig. 3.4 visualizes the cPHA specification

$$\mathcal{CA} = \langle \mathcal{A}_1 \parallel \mathcal{A}_2, \{u_{d1}, u_{d2}, w_{c1}\}, \{w_{c4}\}, \{v_{c1}, v_{c2}, v_{c3}, v_{c4}\}, N \rangle \qquad (3.16)$$

that utilizes the PHA composition of the two-component model that we introduced above in equation (3.12). The I/O variables $\{w_{c2}, w_{c3}\} = \mathbf{w}_c - (\mathbf{u}_c \cup \mathbf{y}_c \cup \mathbf{v}_c)$ are only used to interconnect the automata. We call these I/O variables the *internal I/O variables* of the cPHA. Symbolically solving the set of equations will eliminate these variables from the difference equations so that the difference equations are expressed exclusively in terms of the state variables $\mathbf{x}_c$ and input variables $\mathbf{u}_c$ of the cPHA. As a consequence, the hybrid model captures the state, at a given time-step k, entirely in terms of the mode $\mathbf{x}_{d,k}$ and the set of continuous state variables $\mathbf{x}_{c,k}$, more precisely:

---

[7] For example, sensors can experience different magnitudes of disturbances at different modes.

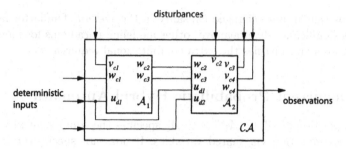

Fig. 3.4. cPHA example with two component automata.

Definition 3.6 (Hybrid state). The *hybrid state* $\mathbf{x}_k$ at time-step k specifies the mode valuation $\mathbf{x}_{d,k}$ of the mode variables $\mathbf{x}_d = \mathbf{x}_{d1}\cup,\ldots,\cup\mathbf{x}_{d\zeta}$ and the continuous state valuation of the continuous state variables $\mathbf{x}_c = \mathbf{x}_{c1}\cup,\ldots,\cup\mathbf{x}_{c\zeta}$ for a cPHA with component automata $\mathcal{A}_1,\ldots,\mathcal{A}_\zeta$.

The separation of I/O variables into input, noise, observed, and internal variables allows us to provide the set of dynamic equations in the usual state-space format, where internal I/O variables are eliminated and the evolution, as well as the observation, is entirely described in terms of the input, state, noise and output variables:

$$\begin{aligned} \mathbf{x}_{c,k+1} &= \mathbf{f}(\mathbf{x}_{c,k}, \mathbf{x}_{d,k+1}, \mathbf{u}_{c,k}, \mathbf{v}_{c,k}) \\ \mathbf{y}_{c,k} &= \mathbf{g}(\mathbf{x}_{c,k}, \mathbf{x}_{d,k}, \mathbf{u}_{c,k}, \mathbf{v}_{c,k}) \, . \end{aligned} \tag{3.17}$$

Although our cPHA/PHA definitions do not constrain the set of equations per se, we do assume that the models of real-world systems lead to 'well formed equations' that permit a symbolic solver to arrive at the state space model form (3.17). Current ongoing research deals with this issue and will provide an extended definition for PHA compatibility that enables a compiler to check whether every possible mode of the composite automaton provides a well formed set of equations. Our current implementation of the hybrid estimation framework uses the extended Kalman filter as the underlying continuous filtering principle. Therefore, we will restrict the model even further, and assume that the disturbance variables $\mathbf{v}_{c,k} = \mathbf{v}_{cx,k} \cup \mathbf{v}_{cy,k}$ describe *white, zero-mean Gaussian noise* that acts *additive* upon the state (state disturbance $\mathbf{v}_{cx,k}$) and output variables (measurement noise $\mathbf{v}_{cy,k}$). More precisely, we will assume the following state space model:

$$\begin{aligned} \mathbf{x}_{c,k+1} &= \mathbf{f}(\mathbf{x}_{c,k}, \mathbf{x}_{d,k+1}, \mathbf{u}_{c,k}) + \mathbf{v}_{cx,k} \\ \mathbf{y}_{c,k} &= \mathbf{g}(\mathbf{x}_{c,k}, \mathbf{x}_{d,k}, \mathbf{u}_{c,k}) + \mathbf{v}_{cy,k} \, . \end{aligned} \tag{3.18}$$

For example, Eqs. (3.15) for the mode

$$\mathbf{x}_{d,k+1} = \begin{bmatrix} m_{11} \\ m_{21} \end{bmatrix} \tag{3.19}$$

lead to the model:

$$\mathbf{x}_{c,k+1} = \begin{bmatrix} 0.7 & -0.2 & 0 \\ 0.3 & 0.8 & 0 \\ 0.3 & -0.2 & 0.5 \end{bmatrix} \mathbf{x}_{c,k} + \begin{bmatrix} 1 \\ 0 \\ 0 \end{bmatrix} \mathbf{u}_{c,k} + \begin{bmatrix} 1 & 0 & 0 & 0 \\ 0 & 1 & 0 & 0 \\ 0 & 0 & 1 & 0 \end{bmatrix} \mathbf{v}_{c,k}$$

$$\mathbf{y}_{c,k} = \begin{bmatrix} 0 & 1 & 1 \end{bmatrix} \mathbf{x}_{c,k} + \begin{bmatrix} 0 & 0 & 0 & 1 \end{bmatrix} \mathbf{v}_{c,k}.$$

(3.20)

The model separates the noise variables $\mathbf{v}_{c,k} = \{v_{c1,k}, v_{c2,k}, v_{c3,k}, v_{c4,k}\}^T$ of the cPHA into the disjoint sets of state disturbances $\mathbf{v}_{cx,k} = \{v_{c1,k}, v_{c2,k}, v_{c3,k}\}$ and measurement noise $\mathbf{v}_{cy,k} = \{v_{c4,k}\}$.

3.4 PHA and cPHA Execution

The *execution* or *trajectory* of a cPHA describes a possible sequence of hybrid states that captures the discrete-time dynamic behavior of the automaton that is interleaved with discrete mode transitions. We adopt the usual hybrid systems view, where the evolution of the system takes place at two distinct time rates: (a) discrete mode changes take place immediately, or at least within ϵ time, while, (b) the continuous evolution over time takes place at the sampling period T_s.

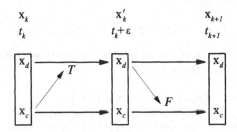

Fig. 3.5. cPHA execution model.

Figure 3.5 visualizes this concept of discrete-time execution for one sampling-period. A hybrid state $\mathbf{x}_k = \{\mathbf{x}_{d,k}, \mathbf{x}_{c,k}\}$, and the input values $\mathbf{u}_{d,k}$ and $\mathbf{u}_{c,k}$, at the time point $t_k = kT_s + t_0$ trigger a mode transition $\mathbf{x}_{d,k} = \mathbf{m}_i \rightarrow \mathbf{m}_j = \mathbf{x}'_{d,k}$ according to the transition specification T of the cPHA. This mode transition leads to a new hybrid state $\mathbf{x}'_k = \{\mathbf{x}'_{d,k}, \mathbf{x}'_{c,k}\}$ with the changed mode $\mathbf{x}'_{d,k} = \mathbf{m}_j$ and the continuous state $\mathbf{x}'_{c,k}$. Time proceeds only infinitesimally (ϵ), therefore, we can assume that the continuous state remains at its previous value $\mathbf{x}'_{c,k} = \mathbf{x}_{c,k}$ or experiences a transition specific reset $\mathbf{x}_{c,k} \xrightarrow{r_i} \mathbf{x}'_{c,k}$. The mode remains constant for the following time period ($T_s - \epsilon$ seconds) and allows the system to evolve according to the discrete-time dynamics $F(\mathbf{m}_j)$. This leads to the *execution* (or hybrid trajectory) of the cPHA

$$\mathbf{x}_k \xrightarrow{\tau_k} \mathbf{x}'_k \xrightarrow{F_k} \mathbf{x}_{k+1} \xrightarrow{\tau_{k+1}} \mathbf{x}'_{k+1} \cdots$$

that specifies a *sequence*, or *ordered set*, of hybrid states and transitions

$$\{\mathbf{x}_0, \tau_0, \mathbf{x}'_0, \mathbf{x}_1, \tau_1, \mathbf{x}'_1, \ldots, \mathbf{x}_k, \tau_k, \mathbf{x}'_k, \mathbf{x}_{k+1}, \tau_{k+1}, \mathbf{x}'_{k+1}, \ldots\} . \qquad (3.21)$$

The execution of a cPHA is subject to observation via the continuous output variables $\mathbf{y}_c$. Observation takes place at the sample time-points $t_k = kT_s + t_0$. The transitions and the intermediate states $\mathbf{x}'_k$ are not directly observed since they denote actions within the interval $(t_k \ \ t_k + \epsilon]$ and states at the time-point $t_k + \epsilon$.

$$\mathbf{x}_k \xrightarrow{\tau_k} \mathbf{x}'_k \xrightarrow{F_k} \mathbf{x}_{k+1} \xrightarrow{\tau_{k+1}} \mathbf{x}'_{k+1} \cdots$$
$$\downarrow \mathcal{O} \qquad\qquad\qquad \downarrow \mathcal{O}$$
$$\mathbf{y}_{c,k} \qquad\qquad\quad \mathbf{y}_{c,k+1}$$

Thus, autonomous and commanded mode changes at the time-point k are observed in terms of their effects onto the continuous state earliest at the following time-step $k + 1$. We call the sequence of observations

$$\{\mathbf{y}_{c,1}, \mathbf{y}_{c,2}, \ldots, \mathbf{y}_{c,k-1}, \mathbf{y}_{c,k}\} =: Y_{c,k} \qquad (3.22)$$

of the cPHA execution (3.21) the *discrete-time trace* and use the somewhat more complicated notation

$$Y_{c,r:k} := \{\mathbf{y}_{c,r}, \mathbf{y}_{c,r+1}, \ldots, \mathbf{y}_{c,k-1}, \mathbf{y}_{c,k}\} \qquad (3.23)$$

whenever the observation starts at a later sample time-point $t_r = rT_s + t_0$.

Analogously, we use the notation

$$\{\mathbf{x}_0, \mathbf{x}_1, \mathbf{x}_2, \ldots, \mathbf{x}_{k-1}, \mathbf{x}_k\} =: X_k \qquad (3.24)$$

to denote the *discrete-time hybrid trajectory* (or *sampled execution*) of the cPHA that starts at the sample time t_0, or

$$\{\mathbf{x}_r, \mathbf{x}_{r+1}, \mathbf{x}_{r+2}, \ldots, \mathbf{x}_{k-1}, \mathbf{x}_k\} =: X_{r:k} \qquad (3.25)$$

whenever the sequence starts at $t_r = rT_s + t_0$. The hybrid trajectory only takes the hybrid states at the sampling-points into account, thus it masks the intermediate states $\mathbf{x}'_k$ after the transition and hides the mode changes up to the next sampling point. This has an important implication for estimation. A mode change in one component at time-step k can trigger a consecutive mode change in another component with a delay of at least one sample. Thus, mode transitions of components at a time-step k are mutually independent, a fact that will help us to formulate an efficient estimation scheme later on.

Hybrid estimation will require us to refer to possible discrete evolutions of the cPHA. We express this in terms of the modes $\mathbf{x}_{d,k}$ and the executed transitions τ_k. We need both specifications, because our transition model does not limit the number of transitions from a mode $\mathbf{m}_i$ to a mode $\mathbf{m}_j$ to one. Several redundant transitions can exist that can specify different reset conditions r_i for the continuous state (see Fig. 3.2). As a consequence, we cannot uniquely determine the discrete evolution based on the mode sequence only and require the additional knowledge about the transitions that were taken in the course of the execution. We abstract a hybrid execution (hybrid trajectory)

$$\{\mathbf{x}_0, \tau_0, \mathbf{x}_0', \mathbf{x}_1, \tau_1, \mathbf{x}_1', \ldots, \mathbf{x}_{k+1}, \tau_{k-1}, \mathbf{x}_{k+1}', \mathbf{x}_k\}$$

into the discrete execution (discrete trajectory)

$$\{\mathbf{x}_{d,0}, \tau_0, \mathbf{x}_{d,1}, \tau_1, \mathbf{x}_{d,2}, \ldots, \mathbf{x}_{d,k-1}, \tau_{k-1}, \mathbf{x}_{d,k}\} =: M_{d,k} \ . \tag{3.26}$$

For clarity and in order to keep statements concise, we will also refer to $M_{d,k}$ as the *mode sequence*. The number of possible mode sequences increases exponentially with the time-steps considered. Nevertheless, due to the finite number of modes and transitions of a cPHA, we can be sure that the number of mode traces is finite for any finite time-point k. This is in contrast to the discrete-time hybrid trajectories, where the real valued state variables imply an infinite number of possible trajectories.

4

Hybrid Estimation

Whenever we automate a complex physical artifact it is essential to know its operational state at each time. However, it is not always possible or desirable to measure all physical entities that determine the state of the system. As a consequence, we have to *estimate* the state, given some (noisy) measurements and the model of the system under investigation. More precisely:

> **Hybrid Estimation Problem:** Given a cPHA model $\mathcal{CA}$ of the system under investigation, the sequences (or discrete-time traces) of observations $Y_{c,k} = \{\mathbf{y}_{c,1}, \dots \mathbf{y}_{c,k}\}$ and the control inputs $U_k = \{\mathbf{u}_0, \mathbf{u}_1, \dots \mathbf{u}_k\}$, estimate the hybrid state $\mathbf{x}_k$ at time-step k, that is comprised of the mode $\mathbf{x}_{d,k}$ and the continuous state $\mathbf{x}_{c,k}$.

Because we model the disturbances and the non-deterministic mode changes probabilistically, we won't obtain an estimate that consists of a particular mode and a crisp valuation for the continuous state. We will obtain an estimate in the form of *distributions* among the possible modes $\mathcal{X}_d$ and among the continuous state space $\mathbb{R}^{n_x}$. We shall see in the next section that hybrid estimation leads to a set of *discrete-time trajectory hypotheses* with varying likelihood. This set of hypotheses, as a whole, defines the overall distributions for the mode and the continuous state. We will develop hybrid estimation incrementally, starting with reviewing traditional estimation, then presenting hybrid estimation for systems that do not exhibit mode changes, and finally, introducing hybrid estimation for systems that can evolve both continuously and discretely through commanded or autonomous mode changes.

4.1 Traditional Estimation

Let us recall traditional estimation, in particular Kalman and extended Kalman filtering prior delving into the theory of hybrid estimation. A Kalman filter [61, 42, 10, 45] utilizes a stochastic linear model of the system under investigation and estimates, in some optimal sense, the physical entities that determine the state of a system. More specifically, let us assume that the system under investigation can be modeled as a discrete-time linear time-invariant (LTI) model of the form:[1]

$$\begin{aligned}
\mathbf{x}_{c,k+1} &= \mathbf{A}\mathbf{x}_{c,k} + \mathbf{B}\mathbf{u}_{c,k} + \mathbf{v}_{cx,k} \\
\mathbf{y}_{c,k} &= \mathbf{C}\mathbf{x}_{c,k} + \mathbf{D}\mathbf{u}_{c,k} + \mathbf{v}_{cy,k} .
\end{aligned} \tag{4.1}$$

The continuous state $\mathbf{x}_c$ and the measurement $\mathbf{y}_c$ are subject to state disturbances $\mathbf{v}_{cx}$ and measurement noise $\mathbf{v}_{cy}$, respectively. The control input $\mathbf{u}_c$, however, is known without error. The initial state of the system is described in terms of a Gaussian distribution $p_{c,0}$ with the mean $\hat{\mathbf{x}}_{c,0}$ and the covariance matrix $\mathbf{P}_0$

$$E\{\mathbf{x}_{c,0}\} = \hat{\mathbf{x}}_{c,0}, \quad E\{(\mathbf{x}_{c,0} - \hat{\mathbf{x}}_{c,0})(\mathbf{x}_{c,0} - \hat{\mathbf{x}}_{c,0})^T\} = \mathbf{P}_0 . \tag{4.2}$$

The state disturbance $\mathbf{v}_{cx}$ and the measurement noise $\mathbf{v}_{cy}$ are assumed to be white, zero-mean Gaussian random sequences with the properties

$$E\{\mathbf{v}_{cx,k}\} = \mathbf{0}, \quad E\{\mathbf{v}_{cx,k}\mathbf{v}_{cx,k}{}^T\} =: \mathbf{Q}, \quad E\{\mathbf{v}_{cx,k}\mathbf{v}_{cx,j}{}^T\} = \mathbf{0}, \quad (j \neq k)$$

$$E\{\mathbf{v}_{cy,k}\} = \mathbf{0}, \quad E\{\mathbf{v}_{cy,k}\mathbf{v}_{cy,k}{}^T\} =: \mathbf{R}, \quad E\{\mathbf{v}_{cy,k}\mathbf{v}_{cy,j}{}^T\} = \mathbf{0}, \quad (j \neq k)$$

$$E\{\mathbf{v}_{cy,k}\mathbf{v}_{cx,j}{}^T\} = \mathbf{0}, \quad \forall j, k .$$

$$\tag{4.3}$$

Neither the state disturbance, nor the measurement noise, is correlated with past continuous states

$$E\{\mathbf{x}_{c,k}\mathbf{v}_{cx,\kappa}{}^T\} = \mathbf{0}, \quad k \leq \kappa$$

$$E\{\mathbf{x}_{c,k}\mathbf{v}_{cy,\kappa}{}^T\} = \mathbf{0}, \quad k \leq \kappa . \tag{4.4}$$

The Kalman filter estimates the continuous state $\mathbf{x}_{c,k}$, based on the continuous inputs $\mathbf{u}_{c,0}, \ldots, \mathbf{u}_{c,k}$, the observations $\mathbf{y}_{c,1}, \ldots, \mathbf{y}_{c,k}$, and the initial state distribution $p_{c,0}$. The estimate represents a multi-variate Gaussian distribution $p_{c,k}$ that is expressed in terms of the mean $\hat{\mathbf{x}}_{c,k}$ and the covariance matrix $\mathbf{P}_k$

$$E\{\mathbf{x}_{c,k}\} = \hat{\mathbf{x}}_{c,k}, \quad E\{(\mathbf{x}_{c,k} - \hat{\mathbf{x}}_{c,k})(\mathbf{x}_{c,k} - \hat{\mathbf{x}}_{c,k})^T\} = \mathbf{P}_k . \tag{4.5}$$

Kalman filtering is performed recursively and minimizes the expected value of the squared sum of estimation errors

---

[1] For the purpose of Kalman filter presentation we deal exclusively with a single continuous model. This would not require us to distinguish between hybrid $(\mathbf{x}, \mathbf{u})$, continuous $(\mathbf{x}_c, \mathbf{u}_c, \mathbf{y}_c)$ and discrete $(\mathbf{x}_d, \mathbf{u}_d)$ variables, thus we could omit the subscript c for the variables. However, for compatibility reasons, we do stick to one coherent notation throughout the monograph, although it might seem redundant for this particular section.

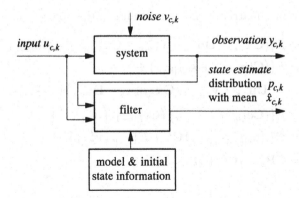

Fig. 4.1. Estimation architecture for standard stochastic filtering.

$$\mathbf{e}_k := \mathbf{x}_{c,k} - \hat{\mathbf{x}}_{c,k} \; . \tag{4.6}$$

This is done by propagating the conditional PDF for the continuous state from one sampling instant to the next, taking into account the system dynamics $(\mathbf{A}, \mathbf{B}, \mathbf{C}, \mathbf{D})$, the input $(\mathbf{u}_c)$, the measurement $(\mathbf{y}_c)$, and the noise characteristics $(\mathbf{Q}, \mathbf{R})$. The recursive deduction of mean $\hat{\mathbf{x}}_{c,k}$ and covariance matrix $\mathbf{P}_k$ is done in two steps:

1. **State extrapolation**

$$\begin{aligned}\hat{\mathbf{x}}_{c,k|k-1} &= \mathbf{A}\hat{\mathbf{x}}_{c,k-1} + \mathbf{B}\mathbf{u}_{c,k-1} \\ \mathbf{P}_{k|k-1} &= \mathbf{A}\mathbf{P}_{k-1}\mathbf{A}^T + \mathbf{Q} \; .\end{aligned} \tag{4.7}$$

We use the subscript $k|k-1$ to indicate that the mean $\hat{\mathbf{x}}_{c,k|k-1}$ and the covariance matrix $\mathbf{P}_{k|k-1}$ represent the *prior estimate* that is based on the past estimate $\hat{\mathbf{x}}_{c,k-1}$, the inputs $\mathbf{u}_{c,0}, \dots, \mathbf{u}_{c,k-1}$, and the measurements $\mathbf{y}_{c,1}, \dots, \mathbf{y}_{c,k-1}$ up to the time-step $k-1$. This one-step ahead prediction leads to a prediction residual, or *innovation* $\mathbf{r}_k$

$$\mathbf{r}_k := \mathbf{y}_{c,k} - \left(\mathbf{C}\hat{\mathbf{x}}_{c,k|k-1} + \mathbf{D}\mathbf{u}_{c,k} \right) \tag{4.8}$$

with the associated covariance matrix $\mathbf{S}_k$

$$\begin{aligned}\mathbf{S}_k &:= E\{\mathbf{r}_k \mathbf{r}_k^T\} \\ &= E\{\left(\mathbf{y}_{c,k} - \mathbf{C}\hat{\mathbf{x}}_{c,k|k-1} - \mathbf{D}\mathbf{u}_{c,k} \right)\left(\mathbf{y}_{c,k} - \mathbf{C}\hat{\mathbf{x}}_{c,k|k-1} - \mathbf{D}\mathbf{u}_{c,k} \right)^T\} \; .\end{aligned}$$

If we define the prior estimation error in analogy to (4.6) as

$$\mathbf{e}_{k|k-1} := \mathbf{x}_{c,k} - \hat{\mathbf{x}}_{c,k|k-1} \; ,$$

we can write the prior covariance matrix $\mathbf{P}_{k|k-1}$ as

$$\begin{aligned}\mathbf{P}_{k|k-1} &= E\{(\mathbf{x}_{c,k} - \hat{\mathbf{x}}_{c,k|k-1})(\mathbf{x}_{c,k} - \hat{\mathbf{x}}_{c,k|k-1})^T\} \\ &= E\{\mathbf{e}_{k|k-1}\mathbf{e}_{k|k-1}^T\} \; .\end{aligned} \tag{4.9}$$

The covariance matrix of the innovation can then be written as

$$
\begin{aligned}
\mathbf{S}_k &= E\{(\mathbf{Cx}_{c,k} + \mathbf{Du}_{c,k} + \mathbf{v}_{cy,k} - \mathbf{C\hat{x}}_{c,k|k-1} - \mathbf{Du}_{c,k})\,(\cdots)^T\} \\
&= E\{(\mathbf{C}(\mathbf{e}_{k|k-1} + \mathbf{\hat{x}}_{c,k|k-1}) + \\
&\quad\quad \mathbf{Du}_{c,k} + \mathbf{v}_{cy,k} - \mathbf{C\hat{x}}_{c,k|k-1} - \mathbf{Du}_{c,k})\,(\cdots)^T\} \\
&= E\{(\mathbf{Ce}_{k|k-1} + \mathbf{v}_{cy,k})\,(\mathbf{Ce}_{k|k-1} + \mathbf{v}_{cy,k})^T\} \\
&= \mathbf{C}E\{\mathbf{e}_{k|k-1}\mathbf{e}_{k|k-1}^T\}\mathbf{C}^T + E\{\mathbf{v}_{cy,k}\mathbf{v}_{cy,k}^T\} \\
&= \mathbf{CP}_{k|k-1}\mathbf{C}^T + \mathbf{R}\,.
\end{aligned}
\tag{4.10}
$$

2. Estimation correction

$$
\mathbf{\hat{x}}_{c,k} = \mathbf{\hat{x}}_{c,k|k-1} + K_k\mathbf{r}_k \tag{4.11}
$$

$$
\mathbf{P}_k = \mathbf{P}_{k|k-1} - K_k\mathbf{S}_kK_k^T \tag{4.12}
$$

provides the *posterior* state estimate with the mean $\mathbf{\hat{x}}_{c,k}$ and the covariance matrix $\mathbf{P}_k$. It is based on the prior estimate and the innovation $\mathbf{r}_k$ with its covariance matrix $\mathbf{S}_k$. The vector K_k denotes the time-variant *Kalman filter gain*:

$$
\begin{aligned}
K_k &= \mathbf{P}_{k|k-1}\mathbf{C}^T\left(\mathbf{CP}_{k|k-1}\mathbf{C}^T + \mathbf{R}\right)^{-1} \\
&= \mathbf{P}_{k|k-1}\mathbf{C}^T\mathbf{S}_k^{-1}\,.
\end{aligned}
\tag{4.13}
$$

Many estimation problems deal with systems that are nonlinear. This prevents one from directly applying the Kalman filter in its original form (4.7)-(4.13). Nevertheless, the success of the Kalman filter, or linear control theory in general, grounds upon the fact that many real world systems can be captured by nonlinear models that can be well approximated by linear models for small perturbations of the state variables. The *extended Kalman filter* [76, 42] utilizes this fact and linearizes the nonlinear model as a (first-order) Taylor series along the expected trajectory. More precisely, let us assume the discrete-time non-linear model

$$
\mathbf{x}_{c,k+1} = \mathbf{f}(\mathbf{x}_{c,k}, \mathbf{u}_{c,k}) + \mathbf{v}_{cx,k} \tag{4.14}
$$

$$
\mathbf{y}_{c,k} = \mathbf{g}(\mathbf{x}_{c,k}, \mathbf{u}_{c,k}) + \mathbf{v}_{cy,k}\,, \tag{4.15}
$$

where $\mathbf{f}$ and $\mathbf{g}$ denote non-linear vector functions of the continuous state and the continuous input. The state disturbance $\mathbf{v}_{cx,k}$ and the measurement noise $\mathbf{v}_{cy,k}$ are, as in the linear case, assumed to be white, zero-mean Gaussian random sequences with the properties given in (4.3)-(4.4).

The non-linear system equations (4.14)-(4.15) imply a non-Gaussian distribution for the state estimate $p_{c,k}$. Nevertheless, given the assumption that the system equations can be adequately approximated by a linear system for small perturbations, it is reasonable to assume that a single multi-variate Gaussian distribution with the mean $\hat{\mathbf{x}}_c$ and the covariance matrix $\mathbf{P}$ represents a proper approximation. The extended Kalman filter performs the approximative filter operation in the following two steps:

1. **State extrapolation**

$$\hat{\mathbf{x}}_{c,k|k-1} = \mathbf{f}(\hat{\mathbf{x}}_{c,k-1}, \mathbf{u}_{c,k-1}) \tag{4.16}$$

$$\mathbf{A}_{k-1} = \left.\frac{\partial \mathbf{f}}{\partial \mathbf{x}}\right|_{\hat{\mathbf{x}}_{c,k-1}, \mathbf{u}_{c,k-1}} \tag{4.17}$$

$$\mathbf{P}_{k|k-1} = \mathbf{A}_{k-1}\mathbf{P}_{k-1}\mathbf{A}_{k-1}^T + \mathbf{Q}. \tag{4.18}$$

This one-step ahead prediction leads to an innovation $\mathbf{r}_k$ with covariance matrix $\mathbf{S}_k$

$$\mathbf{r}_k = \mathbf{y}_{c,k} - \mathbf{g}(\hat{\mathbf{x}}_{c,k|k-1}, \mathbf{u}_{c,k}) \tag{4.19}$$

$$\mathbf{C}_k = \left.\frac{\partial \mathbf{g}}{\partial \mathbf{x}}\right|_{\hat{\mathbf{x}}_{c,k|k-1}, \mathbf{u}_{c,k}} \tag{4.20}$$

$$\mathbf{S}_k = \mathbf{C}_k\mathbf{P}_{k|k-1}\mathbf{C}_k^T + \mathbf{R}. \tag{4.21}$$

2. **Estimation correction**

$$K_k = \mathbf{P}_{k|k-1}\mathbf{C}_k^T\mathbf{S}_k^{-1} \tag{4.22}$$

$$\hat{\mathbf{x}}_{c,k} = \hat{\mathbf{x}}_{c,k|k-1} + K_k\mathbf{r}_k \tag{4.23}$$

$$\mathbf{P}_k = \mathbf{P}_{k|k-1} - K_k\mathbf{S}_kK_k^T. \tag{4.24}$$

Whenever the approximation of the non-linear system in terms of a first-order Taylor series is insufficient, one could utilize higher-order Taylor-series expansions that lead to higher-order extended Kalman filter [13], apply the unscented Kalman filter [60], or use other filtering techniques that were developed for non-linear systems with non-Gaussian distributions, e.g. the versatile *particle filter* [66, 35].

4.2 Hybrid Estimation (Non-switching Case)

Kalman and extended Kalman filtering, presented so far, utilize a *single* dynamic model of the system under investigation and provide a multi-variate distribution p_c as an estimate for the continuous state (Fig. 4.1).

Our aim, however, is different. We intend to estimate the state of a system that can exhibit one out of several modes of operation. As a consequence, we do have a hybrid model, where the system is described in terms of the modes $\mathbf{m}_j \in \{\mathbf{m}_1, \ldots, \mathbf{m}_l\}$ with an associated discrete-time dynamic model

$$\mathbf{x}_{c,k+1} = \mathbf{f}(\mathbf{x}_{c,k}, \mathbf{m}_j, \mathbf{u}_{c,k}) + \mathbf{v}_{c\,x,k} \qquad (4.25)$$

$$\mathbf{y}_{c,k} = \mathbf{g}(\mathbf{x}_{c,k}, \mathbf{m}_j, \mathbf{u}_{c,k}) + \mathbf{v}_{c\,y,k} \; . \qquad (4.26)$$

Generally, we cannot observe the mode of operation directly, so that we have to extend the estimation task to provide both, the estimate for the continuous state, and an estimate for the mode of operation. For the moment, let us assume that the system constantly operates at a particular mode, thus, it cannot switch between the modes during operation (we will relax this assumption later). As an example, one could think of a model where system parameters can take on particular values, but their specific valuation during operation is unknown. We could instantiate a *bank of filters*, one for each mode hypothesis, and operate them concurrently. The filter with the 'best' state estimate identifies the mode of operation, and as a consequence, the specific system parameters that were sought for. This is the principle of *non-switching* or *static multiple-model* estimation [92, 13]. Figure 4.2 illustrates the corresponding hybrid estimation architecture.

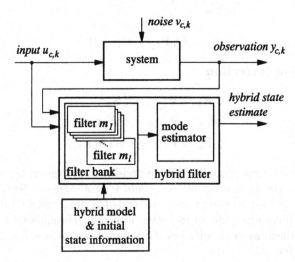

Fig. 4.2. Hybrid estimation architecture for a non-switching hybrid system.

The hybrid estimator is composed of two components. The first component, a (extended Kalman) filter bank, maintains the continuous estimates, one for each mode. The second component, the mode estimator, calculates a conditional probability for each mode, and selects and/or merges the estimation hypotheses to obtain the mode and the overall continuous estimate for the system under investigation. This provides a probability distribution among the modes, as well as the usual continuous estimate.

The mode estimator works as follows. Let $Y_{c,k}$ denote the sequence of continuous measurements $Y_{c,k} = \{\mathbf{y}_{c,1}, \ldots \mathbf{y}_{c,k}\}$. For clarity of the following derivation, and without loss of generality, let us assume that we do not impose any input signal, thus, $\mathbf{u}_{c,i} = 0$ for all $i = 0, \ldots, k$. The mode estimator determines, at each time-step k, the conditional probability $P(\mathbf{m}_j | Y_{c,k})$ that a mode $\mathbf{m}_j$ is the correct mode of operation, given the continuous measurements $Y_{c,k}$. It does so for all modes $\mathbf{m}_j \in \mathcal{X}_d$ and provides a probability (mass) distribution b_k over the set of l possible *mode hypotheses*

$$b_k(\mathbf{m}_j) := P(\mathbf{m}_j | Y_{c,k}) . \tag{4.27}$$

The mode estimator calculates the conditional probabilities recursively, as follows. Let us separate the measurement sequence $Y_{c,k}$ into the leading (fringe) measurement $\mathbf{y}_{c,k}$ and the remaining sequence $Y_{c,k-1}$

$$\begin{aligned} b_k(\mathbf{m}_j) &= P(\mathbf{m}_j | Y_{c,k}) \\ &= P(\mathbf{m}_j | \mathbf{y}_{c,k}, Y_{c,k-1}) . \end{aligned}$$

We then apply Bayes' Rule and calculate $b_k(\mathbf{m}_j)$ as

$$\begin{aligned} b_k(\mathbf{m}_j) &= \frac{1}{c}\, p(\mathbf{y}_{c,k} | \mathbf{m}_j, Y_{c,k-1})\, P(\mathbf{m}_j | Y_{c,k-1}) \\ &= \frac{p(\mathbf{y}_{c,k} | \mathbf{m}_j, Y_{c,k-1})\, b_{k-1}(\mathbf{m}_j)}{\sum_{i=1}^{l} p(\mathbf{y}_{c,k} | \mathbf{m}_i, Y_{c,k-1})\, b_{k-1}(\mathbf{m}_i)}, \quad j = 1, \ldots, l . \tag{4.29} \end{aligned}$$

The underlying deviation can be outlined as follows. Bayes' Rule for a conditional probability $P(A|Y,C)$, where A and C are events for discrete random variables and Y is a continuous random variable, can be written as follows:

$$\begin{aligned} P(A|Y = y, C) &\approx P(A | y \leq Y \leq y + \delta, C) \\ &= \frac{P(y \leq Y \leq y + \delta | A, C)\, P(A|C)}{P(y \leq Y < y + \delta | C)} \\ &\approx \frac{p(y|A,C)\, \delta\, P(A|C)}{p(y|C)\, \delta} . \end{aligned}$$

If we take the limit $\delta \to 0$ and apply the total probability theorem, we obtain the variant of the Bayes' Rule that was used above:

$$P(A_j | y, C) = \frac{p(y|A_j, C)\, P(A_j|C)}{\sum_i p(y|A_i, C)\, P(A_i|C)} .$$

Equation (4.29) establishes the recursive procedure for the mode probability update, what remains to be found is the conditional PDF $p(\mathbf{y}_{c,k} | \mathbf{m}_j, Y_{c,k-1})$. Given that the system is at the mode $\mathbf{m}_j$ and the measurements are known up to time-step $k - 1$ ($Y_{c,k-1} = \{\mathbf{y}_{c,1}, \ldots, \mathbf{y}_{c,k-1}\}$), an associated filter would provide the prior estimate $p_{c,k|k-1}^{(j)}$ with mean $\hat{\mathbf{x}}_{c,k|k-1}^{(j)}$ and covariance matrix $\mathbf{P}_{k|k-1}^{(j)}$ (the superscript index j in parentheses denotes the *mode hypothesis* index, for which the estimate was deduced). As a consequence, we can replace the conditional PDF $p(\mathbf{y}_{c,k}|\mathbf{m}_j, Y_{c,k-1})$ with $p(\mathbf{y}_{c,k}|\hat{\mathbf{x}}^{(j)}_{c,k|k-1})$. The estimate $\hat{\mathbf{x}}^{(j)}_{c,k|k-1}$ for the continuous state would lead to an prior estimate of the output with mean

$$\hat{\mathbf{y}}^{(j)}_{c,k|k-1} = \mathbf{g}\left(\hat{\mathbf{x}}^{(j)}_{c,k|k-1}, \mathbf{m}_j\right) \tag{4.30}$$

and covariance matrix

$$E\left\{\left(\mathbf{y}_{c,k} - \hat{\mathbf{y}}^{(j)}_{c,k|k-1}\right)\left(\mathbf{y}_{c,k} - \hat{\mathbf{y}}^{(j)}_{c,k|k-1}\right)^T\right\} = E\left\{\mathbf{r}_k\mathbf{r}_k^T\right\} = \mathbf{S}_k , \tag{4.31}$$

where the innovation $\mathbf{r}_k$ of the filter for the mode hypothesis $\mathbf{m}_j$ is given by

$$\mathbf{r}_k = \mathbf{y}_{c,k} - \hat{\mathbf{y}}^{(j)}_{c,k|k-1} . \tag{4.32}$$

Therefore, we can obtain the value of the PDF $p(\mathbf{y}_{c,k}|\mathbf{m}_j, Y_{c,k-1})$ from the associated multi-variate Gaussian distribution of the innovation $\mathbf{r}_k$ for hypothesis $\mathbf{m}_j$:

$$p(\mathbf{y}_{c,k}|\hat{\mathbf{x}}^{(j)}_{c,k-1}) = \frac{1}{|2\pi\mathbf{S}_k|^{1/2}} e^{-0.5\mathbf{r}_k^T\mathbf{S}_k^{-1}\mathbf{r}_k} . \tag{4.33}$$

This value can be easily determined in the course of filtering and provides the *measure of likelihood* that is key for the calculation of the conditional probabilities $b_k(\mathbf{m}_j)$.

The distribution b_k can be used to identify the most likely mode $\mathbf{m}_\nu$ $(b_k(\mathbf{m}_\nu) \geq b_k(\mathbf{m}_j)$, for all $j = 1, \ldots, l)$ with its associated continuous (mode-conditioned) estimate $p^{(\nu)}_{c,k}$ that is expressed in terms of the mean and covariance matrix

$$\hat{\mathbf{x}}^{(\nu)}_{c,k}, \quad \mathbf{P}^{(\nu)}_k . \tag{4.34}$$

The combination of the mode-conditioned estimates according to b_k provides the overall continuous estimate with mean

$$\hat{\mathbf{x}}_{c,k} = \sum_{i=1}^{l} b_k(\mathbf{m}_i) \hat{\mathbf{x}}^{(i)}_{c,k} \tag{4.35}$$

and covariance matrix

$$\mathbf{P}_k = \sum_{i=1}^{l} b_k(\mathbf{m}_i) \left[\mathbf{P}^{(i)}_k + (\hat{\mathbf{x}}^{(i)}_{c,k} - \hat{\mathbf{x}}_{c,k})(\hat{\mathbf{x}}^{(i)}_{c,k} - \hat{\mathbf{x}}_{c,k})^T\right] . \tag{4.36}$$

4.3 Full Hypothesis Hybrid Estimation

We now extend the estimation problem and allow switching among the modes according to our cPHA modeling framework. This complicates the hybrid estimation task significantly. Instead of considering the set of possible mode hypotheses, we now have to track the *possible discrete trajectories* or *mode sequences*

$$M_{d,k} = \{\mathbf{x}_{d,0}, \tau_0, \mathbf{x}_{d,1}, \tau_1, \mathbf{x}_{d,2}, \ldots, \mathbf{x}_{d,k-1}, \tau_{k-1}, \mathbf{x}_{d,k}\}$$

and obtain their associated *discrete-time trajectory estimates*

$$\hat{X}_k = \{\hat{\mathbf{x}}_0, \hat{\mathbf{x}}_1, \ldots, \hat{\mathbf{x}}_{k-1}, \hat{\mathbf{x}}_k\} \, . \tag{4.37}$$

Since we are dealing with a *set of trajectory hypotheses*, we will use a superscript index in parentheses, e.g. $M_{d,k}^{(j)}$, to refer to the j'th *trajectory hypothesis*

$$M_{d,k}^{(j)} = \{\mathbf{x}_{d,0}^{(\gamma)}, \tau_0^{(\delta)}, \mathbf{x}_{d,1}^{(\delta)} \ldots, \mathbf{x}_{d,k-1}^{(i)}, \tau_{k-1}^{(j)}, \mathbf{x}_{d,k}^{(j)}\} \, .$$

Hybrid estimation maintains a hybrid discrete-time trajectory estimate

$$\hat{X}_k^{(j)} = \{\hat{\mathbf{x}}_0^{(\gamma)}, \hat{\mathbf{x}}_1^{(\delta)}, \ldots, \hat{\mathbf{x}}_{k-1}^{(i)}, \hat{\mathbf{x}}_k^{(j)}\}$$

for every hypothesis $M_{d,k}^{(j)}$. The estimate tracks the hypothesis up to the time-step k and defines the hybrid state estimate $\hat{\mathbf{x}}_k^{(j)}$ at its fringe. This *fringe estimate* quantifies the mode $\hat{\mathbf{x}}_{d,k}^{(j)} = \mathbf{x}_{d,k}^{(j)}$, and the continuous state estimate in terms of a PDF $p_{c,k}^{(j)}$ with the mean $\hat{\mathbf{x}}_{c,k}^{(j)}$ as the tuple

$$\hat{\mathbf{x}}_k^{(j)} := \langle \hat{\mathbf{x}}_{d,k}^{(j)}, p_{c,k}^{(j)} \rangle \, . \tag{4.38}$$

Hybrid estimation tracks the finite set of trajectory hypotheses $M_{d,k}^{(j)}$ incrementally in terms of their associated estimates $\hat{X}_k^{(j)}$. Estimation starts with a set of initial estimates $\{\hat{X}_0^{(1)}, \ldots, \hat{X}_0^{(\lambda_0)}\}$ that are drawn from the initial state information X_0 of the cPHA model. Hybrid estimation determines for all trajectory estimates $\hat{X}_{k-1}^{(i)} = \{\hat{\mathbf{x}}_0^{(\gamma)}, \ldots, \hat{\mathbf{x}}_{k-1}^{(i)}\}$ all possible transitions $\hat{\mathbf{x}}_{d,k-1}^{(i)} \xrightarrow{\tau_{k-1}^{(j)}} \hat{\mathbf{x}}_{d,k}^{(j)}$ and forms the associated hybrid state estimate $\hat{\mathbf{x}}_k^{(j)}$. Thus, it extends a trajectory estimate $\hat{X}_{k-1}^{(i)}$ with the possible successors $\hat{\mathbf{x}}_k^{(j)}$ and obtains several new trajectory estimates $\hat{X}_k^{(j)} = \{\hat{\mathbf{x}}_0^{(\gamma)}, \ldots, \hat{\mathbf{x}}_{k-1}^{(i)}, \hat{\mathbf{x}}_k^{(j)}\}$ for the consecutive time-step. This operation can be interpreted as building a *full-hypothesis tree* that encodes the estimates for the possible trajectories (mode sequences) that the system can take. Figure 4.3 illustrates this process with a full hypothesis tree at $k = 2$, with a single estimate $\hat{\mathbf{x}}_0^{(1)}$ at the initial time-step $k = 0$.

The estimates $\hat{X}_k^{(j)}$, $j = 1, \ldots, \lambda_k$ of the possible trajectory hypotheses $M_{d,k}^{(j)}$ are ranked according to the conditional probability $P(M_{d,k}^{(j)}|Y_{c,k}, U_k)$ of the hypotheses (U_k denotes the combined sequence of continuous inputs $U_{c,k} = \{\mathbf{u}_{c,0}, \ldots, \mathbf{u}_{c,k}\}$ and discrete (command) inputs $U_{d,k} = \{\mathbf{u}_{d,0}, \ldots, \mathbf{u}_{d,k}\}$). This specifies a probability distribution b_k

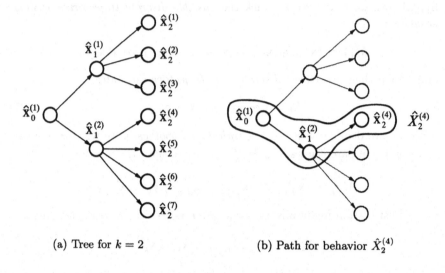

(a) Tree for $k = 2$ (b) Path for behavior $\hat{X}_2^{(4)}$

Fig. 4.3. Full hypothesis tree.

$$b_k(M_{d,k}^{(j)}) := P(M_{d,k}^{(j)}|Y_{c,k}, U_k), \quad j = 1, \ldots, \lambda_k \,, \qquad (4.39)$$

among the set of hypotheses that defines, together with the associated trajectory fringe estimates

$$\{\hat{\mathbf{x}}_k^{(1)}, \ldots, \hat{\mathbf{x}}_k^{(\lambda_k)}\} \,,$$

the hybrid state distribution at the time-step k. In the following we will use the short notation $b_k^{(j)}$ for $b_k(M_{d,k}^{(j)})$.

Trajectory-based estimation with additional probabilistic ranking according to $b_k^{(j)}$ utilizes a hybrid estimation architecture (Fig. 4.4) that is similar to the non-switching hybrid estimation architecture that was introduced above. Again, the full hypothesis hybrid estimator consists of two components, a filter bank that maintains one filter per trajectory hypothesis, and a generalization of the mode estimator that (a) calculates the ranking among the hypotheses and (b) initiates new filters according to the trajectory hypotheses under consideration. Thus, it controls the filter bank that contains a monotonically growing number of dynamic filters. Tracking all possible trajectories that a system can take is almost always intractable, because the number of trajectory hypotheses becomes too large after only a few time steps. Nevertheless, we shall ignore this practicability issue for the moment and deduce the full hypothesis hybrid estimator first. This will contribute to our understanding of the optimal hybrid estimator and its complexity. Once the underlying theory is laid down, we can proceed with suitable sub-optimal hybrid estimation schemes.

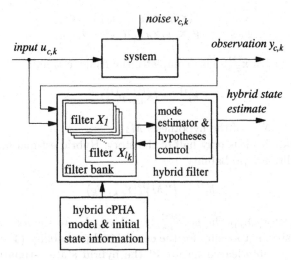

Fig. 4.4. Full hypothesis hybrid estimation architecture.

Let us evaluate the hybrid estimation task for a hypothesis $M_{d,k-1}^{(i)} = \{\ldots, \mathbf{x}_{d,k-1}^{(i)}\}$ with the associated estimate

$$\hat{X}_{k-1}^{(i)} = \{\ldots, \hat{\mathbf{x}}_{k-1}^{(i)}\}.$$

The first step deduces possible mode transitions $\hat{\mathbf{x}}_{d,k-1}^{(i)} \xrightarrow{\tau_{k-1}^{(j)}} \hat{\mathbf{x}}_{d,k}^{(j)}$ and extends trajectory hypothesis with $\tau_{k-1}^{(j)}$ and $\mathbf{x}_{d,k}^{(j)} = \hat{\mathbf{x}}_{d,k}^{(j)}$

$$M_{d,k}^{(j)} = \{\ldots, \mathbf{x}_{d,k-1}^{(i)}, \tau_{k-1}^{(j)}, \mathbf{x}_{d,k}^{(j)}\}. \tag{4.40}$$

The associated hybrid fringe estimate $\hat{\mathbf{x}}_k^{(j)} = \langle \hat{\mathbf{x}}_{d,k}^{(j)}, p_{c,k}^{(j)} \rangle$, which extends the trajectory estimate $\hat{X}_{k-1}^{(i)}$ to

$$\hat{X}_k^{(j)} = \{\ldots, \hat{\mathbf{x}}_{k-1}^{(i)}, \hat{\mathbf{x}}_k^{(j)}\}, \tag{4.41}$$

is deduced as follows: The transition $\tau_{k-1}^{(j)}$ determines the mode $\hat{\mathbf{x}}_{d,k}^{(j)}$ of the hybrid estimate $\hat{\mathbf{x}}_k^{(j)}$, as well as the continuous state estimate $p_{c,k-1}^{'(j)}$ immediately after the transition. The mean $\hat{\mathbf{x}}_{c,k-1}^{'(j)}$ and the covariance matrix $\mathbf{P}_{k-1}^{'(j)}$ of this estimate serves as the initial value for second estimation step. This step performs the extended Kalman filtering task

$$p_{c,k-1}^{'(j)} \rightarrow p_{c,k|k-1}^{(j)} \rightarrow p_{c,k}^{(j)}$$

according to the model

$$\begin{aligned} \mathbf{x}_{c,k+1} &= \mathbf{f}(\mathbf{x}_{c,k}, \mathbf{x}_{d,k+1}, \mathbf{u}_{c,k}) + \mathbf{v}_{cx,k} \\ \mathbf{y}_{c,k} &= \mathbf{g}(\mathbf{x}_{c,k}, \mathbf{x}_{d,k}, \mathbf{u}_{c,k}) + \mathbf{v}_{cy,k} \end{aligned} \tag{4.42}$$

for the mode $\mathbf{x}_{d,k} = \hat{\mathbf{x}}_{d,k}^{(j)}$. This completes the fringe estimate

$$\hat{\mathbf{x}}_k^{(j)} := \langle \hat{\mathbf{x}}_{d,k}^{(j)}, P_{c,k}^{(j)} \rangle \tag{4.43}$$

for the estimate $\hat{X}_k^{(j)}$ of the hypothesis $M_{d,k}^{(j)}$.

The estimate itself is only half of the story. Hybrid estimation also calculates the conditional probability

$$b_k^{(j)} = P(M_{d,k}^{(j)}|Y_{c,k}, U_k) \tag{4.44}$$

for every trajectory hypothesis $M_{d,k}^{(j)}$, $j = 1, \ldots, \lambda_k$. In the following, we derive the recursive relationship for the conditional probability (4.44). This will demonstrate the interleaved nature of the hybrid state estimation and the calculation of the associated conditional probability.

We start off similarly as in the non-switching case and separate the measurement sequence $Y_{c,k}$ into the leading measurement $\mathbf{y}_{c,k}$ and the remaining sequence $Y_{c,k-1}$. This enables us to apply Bayes' Rule and we can calculate $b_k^{(j)}$ as follows:

$$\begin{aligned} b_k^{(j)} &= P(M_{d,k}^{(j)}|\mathbf{y}_{c,k}, Y_{c,k-1}, U_k) \\ &= \frac{1}{c} \, p(\mathbf{y}_{c,k}|M_{d,k}^{(j)}, Y_{c,k-1}, U_k) \, P(M_{d,k}^{(j)}|Y_{c,k-1}, U_k) \,, \end{aligned} \tag{4.45}$$

where c denotes the normalization term

$$c = \sum_{\iota=1}^{\lambda_k} p(\mathbf{y}_{c,k}|M_{d,k}^{(\iota)}, Y_{c,k-1}, U_k) \, P(M_{d,k}^{(\iota)}|Y_{c,k-1}, U_k) \,. \tag{4.46}$$

This establishes the recursive update law for $b_{k-1}^{(i)} \rightarrow b_k^{(j)}$. In order to see this, we have to simplify the two right hand side expressions

$$p(\mathbf{y}_{c,k}|M_{d,k}^{(j)}, Y_{c,k-1}, U_k), \tag{4.47}$$

$$P(M_{d,k}^{(j)}|Y_{c,k-1}, U_k) \tag{4.48}$$

of the equation (4.45) further. Let us start with the conditional PDF (4.47). We use Z_k to denote the combined set of measurement and input sequences

$$Z_k := \{Y_{c,k}, U_k\}.$$

This enables us to rewrite the PDF as

$$\begin{aligned} p(\mathbf{y}_{c,k}|M_{d,k}^{(j)}, Y_{c,k-1}, U_k) &= p(\mathbf{y}_{c,k}|M_{d,k}^{(j)}, Y_{c,k-1}, U_{k-1}, \mathbf{u}_{c,k}, \mathbf{u}_{d,k}) \\ &= p(\mathbf{y}_{c,k}|M_{d,k}^{(j)}, Z_{k-1}, \mathbf{u}_{c,k}, \mathbf{u}_{d,k}) \,. \end{aligned} \tag{4.49}$$

Our cPHA model ensures that the continuous measurement $\mathbf{y}_{c,k}$ at the time-step k is independent of the command input at the same time-step. Thus, we can omit $\mathbf{u}_{d,k}$ in (4.49) and obtain

$$p(\mathbf{y}_{c,k}|M_{d,k}^{(j)}, Z_{k-1}, \mathbf{u}_{c,k}) . \tag{4.50}$$

In order to evaluate the value of this PDF, we have to revisit the filter operation. The first extended Kalman filter step deduces the prior estimate

$$\hat{\mathbf{x}}_{k|k-1}^{(j)} := \langle \hat{\mathbf{x}}_{d,k}^{(j)}, p_{c,k|k-1}^{(j)} \rangle ,$$

given the mode sequence $M_{d,k}^{(j)}$ up to the time-step k, and the inputs as well as the measurements up to the time-step $k-1$. The latter is denoted by Z_{k-1}. This estimate is conditioned on $M_{d,k}^{(j)}$ and Z_{k-1} and enables us to replace $M_{d,k}^{(j)}$ and Z_{k-1} in (4.50) with the prior estimate $\hat{\mathbf{x}}_{k|k-1}^{(j)}$, and we obtain what we call the *hybrid observation function*

$$P_O(\mathbf{y}_{c,k}, \hat{\mathbf{x}}_{k|k-1}^{(j)}, \mathbf{u}_{c,k}) := p(\mathbf{y}_{c,k}|\hat{\mathbf{x}}_{k|k-1}^{(j)}, \mathbf{u}_{c,k}) . \tag{4.51}$$

In the following we will mainly use the short notation

$$P_{O,k}^{(j)} := P_O(\mathbf{y}_{c,k}, \hat{\mathbf{x}}_{k|k-1}^{(j)}, \mathbf{u}_{c,k}) . \tag{4.52}$$

Analogously to the non-switching case ((4.30)-(4.33)) we can draw the value of the hybrid observation function from the multi-variate Gaussian distribution

$$P_{O,k}^{(j)} = \frac{1}{|2\pi \mathbf{S}_k|^{1/2}} e^{-0.5\mathbf{r}_k^T \mathbf{S}_k^{-1}\mathbf{r}_k} . \tag{4.53}$$

Again, $\mathbf{r}_k$, with its associated covariance matrix $\mathbf{S}_k$, denotes the innovation of the j'th hypothesis

$$\mathbf{r}_k = \mathbf{y}_{c,k} - \mathbf{g}\left(\hat{\mathbf{x}}_{c,k|k-1}^{(j)}, \hat{\mathbf{x}}_{d,k}^{(j)}, \mathbf{u}_{c,k}\right) . \tag{4.54}$$

Let us now simplify the second right-hand term $P(M_{d,k}^{(j)}|Y_{c,k-1}, U_k)$ of (4.45). We separate the trajectory hypothesis

$$M_{d,k}^{(j)} = \{\ldots, \mathbf{x}_{d,k-1}^{(i)}, \tau_{k-1}^{(j)}, \mathbf{x}_{d,k}^{(j)}\}$$

into the fringe items $\tau_{k-1}^{(j)}, \mathbf{x}_{d,k}^{(j)}$ and the trajectory prefix $M_{d,k-1}^{(i)} = \{\ldots, \mathbf{x}_{d,k-1}^{(i)}\}$, and utilize the probability law

$$P(A, B|C) = P(A|B, C) \, P(B|C) \tag{4.55}$$

as follows:

$$P(M_{d,k}^{(j)}|Y_{c,k-1},U_k) = P(\tau_{k-1}^{(j)}, \mathbf{x}_{d,k}^{(j)}, M_{d,k-1}^{(i)}|Y_{c,k-1},U_k)$$
$$= P(\tau_{k-1}^{(j)}, \mathbf{x}_{d,k}^{(j)}|M_{d,k-1}^{(i)}, Y_{c,k-1}, U_k)\, P(M_{d,k-1}^{(i)}|Y_{c,k-1},U_k)\,, \quad (4.56)$$

and use the abbreviations P_I and P_{II}

$$P_I := P(\tau_{k-1}^{(j)}, \mathbf{x}_{d,k}^{(j)}|M_{d,k-1}^{(i)}, Y_{c,k-1}, U_k) \quad (4.57)$$

$$P_{II} := P(M_{d,k-1}^{(i)}|Y_{c,k-1},U_k)\,. \quad (4.58)$$

Let us first consider P_I and re-write it as

$$P_I = P(\tau_{k-1}^{(j)}, \mathbf{x}_{d,k}^{(j)}|M_{d,k-1}^{(i)}, Y_{c,k-1}, U_{k-1}, \mathbf{u}_{c,k}, \mathbf{u}_{d,k})$$
$$= P(\tau_{k-1}^{(j)}, \mathbf{x}_{d,k}^{(j)}|M_{d,k-1}^{(i)}, Z_{k-1}, \mathbf{u}_k) \quad (4.59)$$

Again, we replace the mode sequence $M_{d,k-1}^{(i)}$ and the combined measurement/input sequence Z_{k-1} with the optimal estimate $\hat{\mathbf{x}}_{k-1}^{(i)}$, given $M_{d,k-1}^{(i)}$ and Z_{k-1}. Furthermore, we re-apply the probability law (4.55) and obtain:

$$P_I = P(\tau_{k-1}^{(j)}, \mathbf{x}_{d,k}^{(j)}|\hat{\mathbf{x}}_{k-1}^{(i)}, \mathbf{u}_k)$$
$$= P(\mathbf{x}_{d,k}^{(j)}|\tau_{k-1}^{(j)}, \hat{\mathbf{x}}_{k-1}^{(i)}, \mathbf{u}_k)\, P(\tau_{k-1}^{(j)}|\hat{\mathbf{x}}_{k-1}^{(i)}, \mathbf{u}_k)\,. \quad (4.60)$$

These conditional probabilities capture the mode evolution according to the transition

$$\tau_{k-1}^{(j)} = \langle p_{\tau j}, c_j, r_j \rangle\,.$$

The first conditional probability of (4.60) describes the probability of the transition thread to the mode $\mathbf{x}_{d,k}^{(j)}$. This thread probability is conditionally independent of the continuous state $\hat{\mathbf{x}}_{k-1}^{(i)}$ and the input $\mathbf{u}_k$, given that the transition $\tau_{k-1}^{(j)}$ is enabled. This allows us to rewrite the first conditional probability of (4.60) as:

$$P(\mathbf{x}_{d,k}^{(j)}|\tau_{k-1}^{(j)}, \hat{\mathbf{x}}_{k-1}^{(i)}, \mathbf{u}_k) = P(\mathbf{x}_{d,k}^{(j)}|\tau_{k-1}^{(j)})$$
$$= p_{\tau j}(\mathbf{x}_{d,k}^{(j)})\,, \quad (4.61)$$

where $p_{\tau j}(\mathbf{x}_{d,k}^{(j)})$ denotes the value of the transition's probability mass function $p_{\tau j}(\cdot)$ for the transition-target mode $\mathbf{x}_{d,k}^{(j)}$.

The second conditional probability $P(\tau_{k-1}^{(j)}|\hat{\mathbf{x}}_{k-1}^{(i)}, \mathbf{u}_k)$ represents the conditional probability of the transition, given the hybrid state $\hat{\mathbf{x}}_{k-1}^{(i)}$ and the input $\mathbf{u}_k$. We can evaluate this conditional probability in terms of the conditional probability of the transition's guard c_j

$$P(\tau_{k-1}^{(j)}|\hat{\mathbf{x}}_{k-1}^{(i)}, \mathbf{u}_k) = P(c_j|\hat{\mathbf{x}}_{k-1}^{(i)}, \mathbf{u}_k)\,. \quad (4.62)$$

Summing up, guard and thread probability specify P_I as follows

$$P_I = P(\mathbf{x}_{d,k}^{(j)}|\tau_{k-1}^{(j)}) \; P(\tau_{k-1}^{(j)}|\hat{\mathbf{x}}_{k-1}^{(i)}, \mathbf{u}_k) \; . \tag{4.63}$$

This defines what we call the *probabilistic transition function*

$$P_T(\mathbf{x}_{d,k}^{(j)}, \tau_{k-1}^{(j)}, \hat{\mathbf{x}}_{k-1}^{(i)}, \mathbf{u}_k) = P(\mathbf{x}_{d,k}^{(j)}|\tau_{k-1}^{(j)}) \; P(\tau_{k-1}^{(j)}|\hat{\mathbf{x}}_{k-1}^{(i)}, \mathbf{u}_k) \tag{4.64}$$

with its short notation

$$P_{T,k}^{(j)} := P_T(\mathbf{x}_{d,k}^{(j)}, \tau_{k-1}^{(j)}, \hat{\mathbf{x}}_{k-1}^{(i)}, \mathbf{u}_k) \; . \tag{4.65}$$

What remains now, is to simplify the term $P_{II} = P(M_{d,k-1}^{(i)}|Y_{c,k-1}, U_k)$ of (4.56). Again, our cPHA model captures a *causal* system, where the mode at the time-step $k-1$ is independent of the input at the time-step k. As a consequence, we can replace U_k with U_{k-1} in (4.58) and obtain the following expression for P_{II}:

$$
\begin{aligned}
P_{II} &= P(M_{d,k-1}^{(i)}|Y_{c,k-1}, U_k) \\
&= P(M_{d,k-1}^{(i)}|Y_{c,k-1}, U_{k-1}) \\
&= b_{k-1}^{(i)}
\end{aligned}
\tag{4.66}
$$

This relationship establishes the recursive nature of the conditional probability update of full hypothesis hybrid estimation. Hybrid estimation performs the following operation for all trajectory hypotheses $M_{d,k}^{(j)}$, $j = 1, \dots, \lambda_k$ with the associated trajectory estimates $\hat{X}_k^{(j)} = \{\dots, \hat{\mathbf{x}}_{k-1}^{(i)}, \hat{\mathbf{x}}_k^{(j)}\}$

$$
\begin{aligned}
b_k^{(j)} &= P(M_{d,k}^{(j)}|Y_{o,k}, U_k) \\
&= \frac{P_{O,k}^{(j)} \; P_{T,k}^{(j)} \; b_{k-1}^{(i)}}{\sum_{\nu=1}^{\lambda_k} P_{O,k}^{(\nu)} \; P_{T,k}^{(\nu)} \; b_{k-1}^{(i)}}, \quad j = 1, \dots, \lambda_k \; . \; \cdot
\end{aligned}
\tag{4.67}
$$

This recursive calculation of hybrid estimation is very similar to the standard *belief state update* for Hidden Markov Models. Belief state update for HMM is a two-step process that calculates the probability distribution b_k among the (finite) number of l modes $(m_1, \dots, m_l)$ of the HMM at the time-step k. In the first step (prediction), we calculate for every mode m_j the *prior probability* $b_{k|k-1}(m_j)$ of being at this mode, given the transition model of the HMM and the previous belief state b_{k-1}:

$$b_{k|k-1}(m_j) = \sum_{i=1}^{l} P_T(m_j, u_{d,k}, m_i) \; b_{k-1}(m_i), \quad j = 1, \dots, l \; . \tag{4.68}$$

In the second step (refinement), we take the current observation $y_{d,k}$ into account and obtain the *posterior probabilities* $b_k(m_j)$ that define the *belief state* b_k as:

$$b_k(m_j) = \frac{1}{c} \, P_O(y_{d,k}, u_{d,k}, m_j) \, b_{k|k-1}(m_j), \quad j = 1, \ldots, l \, . \tag{4.69}$$

The constant c denotes the normalization factor

$$c = \sum_{i=1}^{l} P_O(y_{d,k}|u_{d,k}, m_i) \, b_{k|k-1}(m_i) \, , \tag{4.70}$$

so that b_k is a probability mass function ($\sum_{i=1}^{l} b_k(m_i) = 1$).

We can write hybrid estimation, as it is given in (4.67), similarly as a recursive two-step process:

Full-Hypothesis Hybrid Estimation:

- The first estimation step extends the trajectory estimates $\hat{X}_{k-1}^{(i)} = \{\ldots, \hat{\mathbf{x}}_{k-1}^{(i)}\}$ of all trajectory hypotheses $M_{d,k}^{(i)}$, $i = 1, \ldots, \lambda_{k-1}$, according to the possible transitions $\hat{\mathbf{x}}_{d,k-1}^{(i)} \xrightarrow{\tau_{k-1}^{(j)}} \hat{\mathbf{x}}_{d,k}^{(j)}$. This determines the modes $\hat{\mathbf{x}}_{d,k}^{(j)}$ for the estimates

$$\hat{X}_k^{(j)} = \{\ldots, \hat{\mathbf{x}}_{k-1}^{(i)}, \hat{\mathbf{x}}_k^{(j)}\}$$

of the extended trajectory hypotheses $M_{d,k}^{(j)}$, $j = 1, \ldots, \lambda_k$, and evaluates the transition-specific change for the continuous estimates

$$p_{c,k-1}^{(i)} \to p_{c,k-1}^{'(j)} \, .$$

It then calculates for every hypothesis $M_{d,k}^{(j)}$ the conditional *prior probability* $b_{k|k-1}^{(j)}$, given the previous posterior probability $b_{k-1}^{(i)}$ and the conditional mode transition probability $P_{T,k}^{(j)}$ of the associated transition $\tau_{k-1}^{(j)}$

$$b_{k|k-1}^{(j)} = P_{T,k}^{(j)} \, b_{k-1}^{(i)}, \quad j = 1, \ldots, \lambda_k \, . \tag{4.71}$$

- The second estimation step takes the current measurement $y_{c,k}$ into account. It performs the extended Kalman filtering that deduces the continuous estimate

$$p_{c,k-1}^{'(j)} \to p_{c,k|k-1}^{(j)} \to p_{c,k}^{(j)}$$

according to the model for the mode $\hat{\mathbf{x}}_{d,k}^{(j)}$. Filtering also provides the probabilistic observation function $P_{\mathcal{O},k}^{(j)}$ for every trajectory estimate $M_{d,k}^{(j)}$. This value is used to obtain the conditional *posterior probabilities* $b_k^{(j)}$

$$b_k^{(j)} = \frac{1}{c} \, P_{\mathcal{O},k}^{(j)} \, b_{k|k-1}^{(j)}, \quad j = 1, \dots, \lambda_k \, , \qquad (4.72)$$

where the constant c denotes the normalization factor

$$c = \sum_{\nu=1}^{\lambda_k} P_{\mathcal{O},k}^{(\nu)} \, b_{k|k-1}^{(\nu)} \, . \qquad (4.73)$$

The major difference between belief state update for Hidden Markov Models and the full hypothesis hybrid estimation is that belief state update merges trajectories that end up in the same mode (summation in (4.68)), whereas hybrid estimation tracks a set of trajectory hypotheses that is exponential in the number of time-steps considered (4.71).

Nevertheless, the conditional probabilities $\{b_k^{(1)}, \dots, b_k^{(\lambda_k)}\}$, together with the hybrid state estimates $\{\hat{\mathbf{x}}_k^{(1)}, \dots, \hat{\mathbf{x}}_k^{(\lambda_k)}\}$ at the fringe of the trajectory estimates $\{\hat{X}_k^{(1)}, \dots, \hat{X}_k^{(\lambda_k)}\}$, *encode* the *hybrid belief state* $b_k(\cdot)$ as follows:

- The belief of being in mode $\mathbf{m}_j$ at time-step k is given by the sum of the conditional probabilities of trajectory hypotheses with a fringe state at the mode $\mathbf{m}_j$:

$$b_k(\mathbf{m}_j) = \sum_{\nu \,|\, \hat{\mathbf{x}}_{d,k}^{(\nu)} = \mathbf{m}_j} b_k^{(\nu)} \, . \qquad (4.74)$$

- The belief of being at the mode $\mathbf{m}_j$ at time step k, with a continuous state within a region $\mathcal{X}_c \subset \mathbb{R}^{n_x}$, is given by:

$$b_k(\mathbf{m}_j, \mathbf{x}_{c,k} \in \mathcal{X}_c) = \sum_{\nu \,|\, \hat{\mathbf{x}}_{d,k}^{(\nu)} = \mathbf{m}_j} \left(b_k^{(\nu)} \int_{\mathbf{x} \in \mathcal{X}_c} p_{c,k}^{(\nu)}(\mathbf{x}) \, d\mathbf{x} \right) \, . \qquad (4.75)$$

Furthermore, we can use the conditional probabilities $b_k^{(j)}$ and the fringe state estimates $p_{c,k}^{(j)}$, with their associated mean $\hat{\mathbf{x}}_{c,k}^{(j)}$ and covariance matrices $\mathbf{P}_k^{(j)}$, to calculate the overall continuous estimate. This is done by mixing the associated PDFs according to their conditional probabilities:

$$p_{c,k} = \sum_{\nu=1}^{\lambda_k} b_k^{(\nu)} p_{c,k}^{(\nu)} \, . \qquad (4.76)$$

This leads to an overall continuous estimate with the mean

$$\hat{\mathbf{x}}_{c,k} = \sum_{\nu=1}^{\lambda_k} b_k^{(\nu)} \hat{\mathbf{x}}_{c,k}^{(\nu)} \tag{4.77}$$

and the covariance matrix

$$\mathbf{P}_k = \sum_{\nu=1}^{\lambda_k} b_k^{(\nu)} \left[\mathbf{P}_k^{(\nu)} + (\hat{\mathbf{x}}_{c,k}^{(\nu)} - \hat{\mathbf{x}}_{c,k})(\hat{\mathbf{x}}_{c,k}^{(\nu)} - \hat{\mathbf{x}}_{c,k})^T \right] . \tag{4.78}$$

The continuous estimate, given the mode $\mathbf{m}_j$, can be obtained by mixing the PDFs of the hypotheses with a fringe state $\hat{\mathbf{x}}_k^{(\nu)}$ at mode $\hat{\mathbf{x}}_{d,k}^{(\nu)} = \mathbf{m}_j$, more precisely:

$$p_{c,k}\big|_{\mathbf{x}_{d,k}=\mathbf{m}_j} = \frac{1}{b_k(\mathbf{m}_j)} \sum_{\nu \,|\, \hat{\mathbf{x}}_{d,k}^{(\nu)}=\mathbf{m}_j} b_k^{(\nu)} p_{c,k}^{(\nu)} . \tag{4.79}$$

Hybrid estimation as presented above utilizes a probabilistic transition function P_T, a probabilistic observation function P_O, and an underlying continuous estimation / filtering mechanism. Filtering and the deduction of the observation function P_O, based on the innovation of the filter's prediction step, was introduced above. What remains now, is to introduce the deduction of the probabilistic *transition function* P_T.

Hybrid transition function

A mode transition $\mathbf{m}_i \rightarrow \mathbf{m}_j$ involves individual transitions $\tau_\eta := \langle p_{\tau\eta}, c_\eta, r_\eta \rangle$, $\tau_\eta \in \mathcal{T}_\nu$ for every component $\mathcal{A}_\nu$ of the cPHA[2]. Given that the automaton component $\mathcal{A}_\nu$ is in mode $\mathbf{x}_{d\nu,k-1} = \mathbf{m}_i$, the conditional probability that it will take a transition to $\mathbf{x}_{d\nu,k-1}' = \mathbf{m}_j$ is the conditional probability that its guard c_η is satisfied, given the continuous state $\mathbf{x}_{c,k-1}$ and the inputs $\mathbf{u}_{c,k-1}, \mathbf{u}_{d,k-1}$, times $p_{\tau\eta}(\mathbf{m}_j)$, the probability of the transition thread that takes the component to the mode $\mathbf{m}_j$, given that the guard c_η is satisfied (i.e. the transition τ_η is enabled). We assumed independence of component transitions in our cPHA model, therefore, we obtain the overall P_T by taking the product of the transition probabilities of the individual components.

For a PHA $\mathcal{A}_\nu$, a guard c_η is a constraint over the continuous state variables $\mathbf{x}_c$, the continuous input $\mathbf{u}_c$, and the discrete command input $\mathbf{u}_d$. A guard c_η is typically of the form (3.2)

$$c_\eta(\mathbf{x}_c, \mathbf{u}_c, \mathbf{u}_d) = c_{xc}(\mathbf{x}_c) \wedge c_{uc}(\mathbf{u}_c) \wedge c_{ud}(\mathbf{u}_d) , \tag{4.80}$$

where $c_{xc}(\cdot)$ and $c_{uc}(\cdot)$ are inequality constraints on the continuous state $\mathbf{x}_c$ and the continuous input $\mathbf{u}_c$, respectively, and $c_{ud}(\cdot)$ is a constraint on the discrete (command) input $\mathbf{u}_d$ in the form of a propositional logic formula. The

---

[2] For symmetry, we also treat the 'non-transition' $\mathbf{m}_j \rightarrow \mathbf{m}_j$ of a PHA $\mathcal{A}_\nu$ as a transition.

constraints are conditionally independent, given the continuous state $\mathbf{x}_{c,k-1}$, and the inputs $\mathbf{u}_{c,k-1}, \mathbf{u}_{d,k-1}$, so that we can write

$$P(c_\eta | \mathbf{x}_{c,k-1}, \mathbf{u}_{c,k-1}, \mathbf{u}_{d,k-1}) = P(c_{xc} | \mathbf{x}_{c,k-1}) P(c_{uc} | \mathbf{u}_{c,k-1}) P(c_{ud} | \mathbf{u}_{d,k-1})$$

and determine the conditional probabilities of the individual constraints separately.

The valuations $\mathbf{u}_{c,k-1}$ and $\mathbf{u}_{d,k-1}$ of the continuous and discrete inputs $\mathbf{u}_c$ and $\mathbf{u}_d$, represent crisp values. This implies that the conditional probabilities $P(c_{uc} | \mathbf{u}_{c,k-1})$ and $P(c_{ud} | \mathbf{u}_{d,k-1})$ have probability 1.0 or 0.0, according to the truth value of the associated guard. The evaluation of $P(c_{xc} | \mathbf{x}_{c,k-1})$, however, is less trivial since we do not know the exact value for the state, but only its estimate. This estimate is given in the form of a multi-variate PDF $p_{c,k-1}$ for the state variables. As a consequence, we have to utilize

$$P(c_{xc} | p_{c,k-1})$$

instead, and calculate the conditional probability by evaluating the volume integral

$$P(c_{xc} | p_{c,k-1}) = \int_{\mathbf{x} \in \mathcal{Q}} p_{c,k-1}(\mathbf{x}) \, d\mathbf{x} \, , \qquad (4.81)$$

where $\mathcal{Q} \subset \mathbb{R}^{n_x}$ denotes the domain for which the guard inequality is satisfied.

For example, consider a continuous guard for a single continuous state variable x_c

$$2.0 \leq x_c < \infty \, . \qquad (4.82)$$

The estimate x_c at the time-step $k-1$ specifies a Gaussian distribution $p_{c,k-1}$ with the mean $\hat{x}_{c,k-1} = 1.0$ and the variance $\sigma^2 = 1.0$. The probability calculation corresponds to determining the black area in Fig. 4.5a and leads to a conditional probability

$$P(c_{xc} | p_{c,k-1}) = \int_2^\infty \frac{1}{\sqrt{2\pi}\sigma} e^{-(x - \hat{x}_{c,k-1})^2 / 2\sigma^2} \, dx = 0.1587 \, . \qquad (4.83)$$

Numerically, this can be done by utilizing the normalized Gaussian *cumulative distribution function (CDF)* $\Phi(\cdot)$

$$P(c_{xc} | p_{c,k-1}) = 1 - \Phi\left(\frac{2 - \hat{x}_{c,k-1}}{\sigma}\right) = 1 - \Phi(1) = 0.1587 \, .$$

The evaluation of the volume integral (4.81) for a general multi-variate constraint c_{xc}, such as

$$-\infty \leq 1 + x_{c1} + x_{c1}^3 - x_{c2} < 0 \, , \qquad (4.84)$$

is non-trivial. As a consequence, we utilize a Monte-Carlo based sampling approach that checks the guard against a sufficiently large set of state samples.

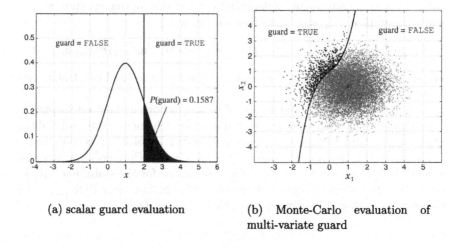

(a) scalar guard evaluation

(b) Monte-Carlo evaluation of multi-variate guard

Fig. 4.5. Guard probability evaluation.

These samples are randomly generated according to the multi-variate distribution $p_{c,k-1}$ of the continuous state estimate. Figure 4.5b visualizes this evaluation for an estimate that is characterized by a Gaussian distribution with the following mean $\hat{\mathbf{x}}_{c,k-1}$ and covariance matrix $\mathbf{P}$:

$$\hat{\mathbf{x}}_{c,k-1} = \begin{bmatrix} 1 \\ 0 \end{bmatrix}, \quad \mathbf{P} = \begin{bmatrix} 1 & 0 \\ 0 & 1 \end{bmatrix}.$$

Out of the 10,000 samples that were taken, 819 satisfy the guard (black samples, to the left of the guard boundary $1 + x_{c1} + x_{c1}^3 - x_{c2} = 0$). This leads to the conditional guard probability

$$P(c_{xc}|p_{c,k-1}) = \frac{819}{10000} = 0.0819.$$

An open research issue is to utilize other approaches to compute $P(c_{xc}|p_{c,k-1})$ more efficiently (see, for example, [94]). This could be done through a combination of restricting and approximating the function $q_{xc}(\mathbf{x}_c)$ of the guard c_{xc}.

Once the transition probabilities

$$P_{T\nu,k}^{(j)} = P(\mathbf{x}_{d\nu,k}^{(j)}, \tau_\eta | \hat{\mathbf{x}}_{k-1}^{(i)}, \mathbf{u}_{k-1})$$
$$= p_{\tau\eta}(\mathbf{x}_{d\nu,k}^{(j)}) \, P(c_\eta | \hat{\mathbf{x}}_{k-1}^{(i)}, \mathbf{u}_{c,k-1}, \mathbf{u}_{d,k-1}) \tag{4.85}$$

for the transition $\tau_\eta = \langle p_{\tau\eta}, c_\eta, r_\eta \rangle$, $\tau_\eta \in \mathcal{T}_\nu$ of all components $\mathcal{A}_\nu$, $\nu = 1, \ldots, \zeta$ are determined, we can calculate the overall probabilistic transition function $P_{T,k}^{(j)}$ for a particular trajectory estimate $\hat{X}_k^{(j)}$ by taking the product of the (independent) component transitions

$$P^{(j)}_{\mathcal{T},k} = \prod_{\nu=1}^{\zeta} P^{(j)}_{\mathcal{T}\nu,k} \, . \tag{4.86}$$

Full hypothesis estimation – epilogue

Hybrid estimation, as introduced above, considers *all possible mode transitions* that can occur during the dynamic evolution of the system. This motivates the name *full hypothesis tree* for the fully grown hypothesis tree that considers estimates up to time-step k. It builds the tree incrementally, starting from an initial set of $\lambda_0 \leq l$ state estimates (l denotes the number of possible modes of the cPHA). Expanding the full tree, which is equivalent to tracking all possible trajectories of a system, is almost always intractable because the number of trajectories becomes too large after only a few time steps. Reconsider the 10 component cPHA example that was mentioned above. The components have in average 5 modes and each mode has in average 3 successor states. This cPHA represents an automaton with $5^{10} \approx 10,000,000$ modes! Hybrid estimation, as formulated above, and a single initial estimate $\hat{\mathbf{x}}_0^{(1)}$ leads to a, worst-case exponentially, increasing number of $\lambda_k = (3^{10})^k$ trajectories that are to be tracked at a time-step k. Figure 4.6 visualizes this blowup for a third hybrid estimation step.

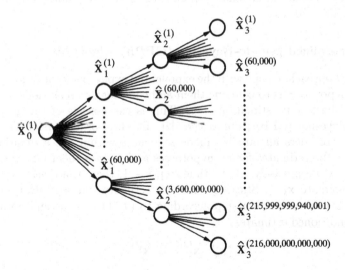

Fig. 4.6. Three-step full hypothesis tree for a (moderately) complex system.

An approach to coping with this exponential explosion is presented in [109], where the individual mode sequences under consideration are averaged. This leads to an optimal prediction of the continuous state alone. Whenever one is interested in on-line estimation of the full hybrid state, that is, continuous state and mode, it is inevitable to use approximate hybrid estimation schemes. These algorithms merge trajectory hypotheses and/or prune unlikely hypotheses, so that the number of hypotheses under consideration stays within a certain limit.

The problem of exponential growth has been addressed by a variety of approximate hybrid estimation schemes, mostly referred to as *multiple-model (MM)* estimation algorithms, in literature. The next section reviews these algorithms and introduces the most prominent one, the interacting multiple-model algorithm (IMM) [20]. This algorithm will later serve as a base-line to our novel hybrid estimation scheme, which we will present afterwards in Section 4.5.

4.4 Multiple-Model Estimation

The *multiple-model (MM)* estimation algorithms are powerful sub-optimal estimation algorithms for the hybrid estimation problem. Their applications are to be mainly found in fields of aeronautics and aerospace, for example for target tracking. Examples for this class of estimation algorithms are the generalized pseudo-Bayesian (GPB) [2], the detection-estimation [97], the residual correlation Kalman filter bank [48], and the interacting multiple-model algorithm (IMM) [20, 13].

The generalized pseudo-Bayesian (GPBn) Algorithm

A simple approach to cope with the exponential explosion of the number of trajectory hypotheses is to combine them and consider different mode sequences only for the last n estimation steps. This is the essence of the generalized pseudo-Bayesian (GPBn) approaches [2]. The first-order version GBP1, for instance, considers all possible mode sequences within one estimation-step, calculates the estimates for all hypotheses and merges them into one hybrid estimate. More precisely, at the time-step $k - 1$, there is a *single, aggregated* hybrid estimate $\hat{\mathbf{x}}_{k-1}$. Starting with this estimate, the algorithm performs a single hybrid estimation step according to (4.71) - (4.73) and obtains the mode-conditioned estimates

$$\{\hat{\mathbf{x}}_k^{(1)}, \hat{\mathbf{x}}_k^{(2)}, \ldots, \hat{\mathbf{x}}_k^{(l)}\}, \tag{4.87}$$

where $\hat{\mathbf{x}}_k^{(j)} = \langle \mathbf{m}_j, p_{c,k}^{(j)} \rangle$ denotes the hybrid state estimate with the mode $\mathbf{m}_j$ and the continuous estimate $p_{c,k}^{(j)}$. l denotes the number of modes, which the hybrid model can take on. The algorithm merges the estimates according to their conditional probabilities $b_k^{(j)}$. This leads to, what we previously called, the overall continuous estimate (4.76)-(4.78). This overall estimate can be expressed in terms of its lumped PDF

$$p_{c,k} = \sum_{\nu=1}^{l} b_k^{(\nu)} p_{c,k}^{(\nu)} \,,$$

or its mean value $\hat{\mathbf{x}}_{c,k}$ and the covariance matrix $\mathbf{P}_k$, as given above in (4.77)-(4.78). In terms of the mode, one can take the *maximum a posterior (MAP)* approach and select the most likely mode $\hat{\mathbf{x}}_{d,k} = \mathbf{m}_\xi$, according to the conditional probabilities b_k of the mode-conditioned estimates (4.87), that is, $b_k^{(\xi)} \geq b_k^{(\nu)}$ for all $\nu = 1,\dots,l$.

In its generalized form, a GPBn algorithm calculates all trajectory hypotheses for n-steps and merges them into a single hybrid estimate. One can view this operation as building, at each time-step k, an n-step full hypotheses tree that originates at an aggregated hybrid estimate $\hat{\mathbf{x}}_{k-n}$. As a consequence, the algorithm requires at most l^n concurrent filtering operations per estimation step.

The Interacting Multiple-Model (IMM) Algorithm

A good trade-off between computational cost and estimation quality is achieved by the *interacting multiple-model* algorithm [20]. IMM provides an estimate with quality similar to GPB2, but only requires l concurrent filters per estimation step (one filter per mode $\mathbf{m}_j \in \mathcal{X}_d$ of the hybrid model). Each filter uses a different combination (mixing) of the previous mode-conditioned estimates

$$\{\hat{\mathbf{x}}_{k-1}^{(1)}, \hat{\mathbf{x}}_{k-1}^{(2)}, \dots, \hat{\mathbf{x}}_{k-1}^{(l)}\} \tag{4.88}$$

as initial value for the estimation at time-step k. Again, the set (4.88) contains one estimate for each mode $\mathbf{m}_j \in \mathcal{X}_d$ of the cPHA model and we assume that the superscript index j of the estimate $\hat{\mathbf{x}}_{k-1}^{(j)}$ directly refers to the mode $\mathbf{m}_j$ of the estimate $(\hat{\mathbf{x}}_{d,k-1}^{(j)} = \mathbf{m}_j)$.

The algorithm can be interpreted as a two-step hidden Markov model (HMM) style belief-state update that determines the conditional probability distribution for the set of modes $b_k^{(j)} = b_k(\mathbf{m}_j)$, $j = 1,\dots,l$, together with an associated continuous filtering operation. The first step calculates the *prior* probability for being at a mode $\mathbf{m}_j$ as

$$b_{k|k-1}^{(j)} = \sum_{i=1}^{l} P(\mathbf{m}_j | \hat{\mathbf{x}}_{k-1}^{(i)}, \mathbf{u}_{k-1})\, b_{k-1}^{(i)}, \; j = 1,\dots,l\,. \tag{4.89}$$

$P(\mathbf{m}_j | \hat{\mathbf{x}}_{k-1}^{(i)}, \mathbf{u}_{k-1})$ denotes the conditional *transition probability* from the mode $\mathbf{m}_i$ to the mode $\mathbf{m}_j$, since the hybrid estimate $\hat{\mathbf{x}}_{k-1}^{(i)}$ denotes an estimate with the mode $\mathbf{m}_i$. This step also provides the l^2 *IMM mixing probabilities*

$$\mu_{ij} = \frac{P(\mathbf{m}_j | \hat{\mathbf{x}}_{k-1}^{(i)}, \mathbf{u}_{k-1})\, b_{k-1}^{(i)}}{b_{k|k-1}^{(j)}} \,, \tag{4.90}$$

which specify the "level of interaction" among the modes. The following mixing operation determines the initial condition for the mode-conditioned filters

$$\bar{p}_{c,k-1}^{(j)} = \sum_{i=1}^{l} p_{c,k-1}^{(i)} \mu_{ij}, \ j = 1,\ldots,l \, . \tag{4.91}$$

Of course, an (extended) Kalman filter uses a mean and a covariance matrix to express a PDF estimate. Thus, one utilizes the following mixed mean $\bar{\mathbf{x}}_{c,k-1}^{(j)}$ and covariance matrices $\bar{\mathbf{P}}_{k-1}^{(j)}$ as initial condition for the mode-conditioned filters:

$$\bar{\mathbf{x}}_{c,k-1}^{(j)} = \sum_{i=1}^{l} \hat{\mathbf{x}}_{c,k-1}^{(i)} \mu_{ij}, \tag{4.92}$$

$$\bar{\mathbf{P}}_{k-1}^{(j)} = \sum_{i=1}^{l} \left[\mathbf{P}_{k-1}^{(i)} - (\hat{\mathbf{x}}_{c,k-1}^{(i)} - \bar{\mathbf{x}}_{c,k-1}^{(i)})(\hat{\mathbf{x}}_{c,k-1}^{(i)} - \bar{\mathbf{x}}_{c,k-1}^{(i)})^T \right] \mu_{ij} \, . \tag{4.93}$$

The second step of the IMM algorithm applies one extended Kalman filter per mode $\mathbf{m}_j$. The filters use the mixed mode conditioned estimates $\{\bar{\mathbf{x}}_{c,k-1}^{(j)}, \bar{\mathbf{P}}_{k-1}^{(j)}\}$ as initial condition and provide the updated set of mode-conditioned estimates

$$\{\hat{\mathbf{x}}_k^{(1)}, \hat{\mathbf{x}}_k^{(2)}, \ldots, \hat{\mathbf{x}}_k^{(l)}\} \, , \tag{4.94}$$

together with the associated hybrid observation functions $Po_k^{(j)}$, $j = 1,\ldots,l$, as described above in (4.51)-(4.54). These hybrid observation functions are used to calculate the *posterior* probabilities for the mode-conditioned estimates at time-step k as

$$b_k^{(j)} = \frac{Po_k^{(j)} b_{k|k-1}^{(j)}}{\sum_{\nu=1}^{l} Po_k^{(\nu)} b_{k|k-1}^{(\nu)}} \, . \tag{4.95}$$

Finally, the IMM algorithm combines the mode-conditioned estimates and provides the overall continuous estimate

$$p_{c,k} = \sum_{i=1}^{l} p_{c,k}^{(i)} b_k^{(i)} \tag{4.96}$$

as estimation output. The most likely mode $\mathbf{m}_\xi$ according to $b_k^{(\xi)} \geq b_k^{(\nu)}$, $\nu = 1,\ldots,l$ can be used to specify the mode for the overall hybrid estimate

$$\hat{\mathbf{x}}_k = \langle \mathbf{m}_\nu, p_{c,k} \rangle \, . \tag{4.97}$$

Adaptive Multiple-Model Estimation

The IMM algorithm is an example for a sub-optimal hybrid estimation scheme that achieves tractability by limiting the number of hypotheses that are checked in the course of estimation. It requires as many filters as there are modes in the hybrid model. In a real-world example, where many components can have several modes of operation and can fail and deteriorate in many ways, this number of modes can still be very large. Recall the example given above – 10 components with 5 modes each. Only a small portion of the mode-space (10,000,000 modes) will be relevant at each time-step k. As a consequence, the evaluation of all mode hypotheses will lead to unnecessarily high computational cost and deteriorate the estimation result as too many irrelevant filters compete against the small portion of relevant filters [68].

Adaptive MM-estimation was proposed as a possible solution for this dilemma [68, 69, 67]. This estimation scheme, that is also called *variable-structure MM-estimation* in literature [68], adapts the mode-set to a subset of modes, that are most likely at a given time-point. The various adaptive MM-estimation methods cited above differ by their selection scheme that determines the appropriate subset of modes $\mathcal{M}_{(k)} \subset \mathcal{X}_d$. The digraph switching algorithm presented in [68], for instance, adapts the mode-set according to the transition graph of the model and switches among groups of closely related modes. The mode transition graph for the simple flow regulator model with three nominal modes in Fig. 4.7, for example, would imply three mode-sets: {closed, partly-open }, {closed, partly-open, fully-open} and {partly-open, fully-open}.

Fig. 4.7. Mode transition graph for a simple flow regulator model.

Adaptive MM-estimation works in two phases. The first *mode-set adaption phase* uses a statistical test to choose a suitable mode-set $\mathcal{M}_{(k)}$. In the second phase, estimation proceeds by evaluating all modes $\mathbf{m}_i$ within the set $\mathcal{M}_{(k)}$. This clear separation of mode-set adaption and estimation has the advantage that various mode-set adaption schemes can be used together with a standard MM-estimation algorithm. A major drawback, however, is that the algorithm does not make use of posterior information that is gathered in the course of estimation. Furthermore, the individual model sets can still be very large for models that capture real-world applications. Reconsider the ten component example above and assume that each mode of a component has in average 2 successor modes (one nominal and one fault mode). An adaptive

MM-estimation method considers the current mode, and its two successors for each component. This leads to $3^{10} \approx 60,000$ modes – still unrealistically large for on-line estimation.

4.5 Focused Hybrid Estimation

Some monitoring and diagnosis systems that build upon the discrete model-based reasoning paradigm, for example the Livingstone and Titan systems [101], utilize the concept of Hidden Markov Model (HMM) style *belief-state update*. They reformulate a discrete, constraint-based estimation problem as a multi-attribute utility problem in the style of HMM belief-state update, and employ advanced search techniques [49, 106] from the toolkit of model-based reasoning to focus estimation onto the *most likely* estimates. Such an approach is successful because a small subset of the possible hypotheses, that is, the set of the most likely ones, is typically sufficient to cover most of the probability space. For hybrid estimation, we utilize the similarities of full-hypothesis hybrid estimation (4.71)-(4.72) and the HMM style *belief-state update* (4.68)-(4.69) and re-frame the estimation task as search. One can interpret this operation as carefully exploring the full hypothesis tree to focus onto the leading set of trajectory hypotheses. We first demonstrated this principle in [55], where we introduced one specific variant of the class of search-based hybrid estimation algorithms that we present below.

Let us start with reconsidering the full hypothesis estimation problem visualized as a full hypothesis tree of Fig. 4.8a. It is reasonable to assume that only a small subset of the 216,000,000,000,000 hypotheses at time-step 3 is relevant to describe the hybrid state of the system at this time-step. We could also say that a small subset of the set of possible hypotheses covers most of the probability space in terms of b_k. The majority of unlikely hypotheses have little impact onto the overall estimation result due to their negligible likelihood $b_k^{(j)}$. However, calculating these hypotheses imposes most of the computational burden that is encountered in the course of hybrid estimation. It is therefore apparent to avoid their deduction as much as possible, and to search the hypothesis tree for the set of most likely hypotheses only.

Key for real-time operation is to provide a search-based hybrid estimation algorithm that incrementally returns estimation hypotheses, starting with the most likely one. Once the most likely hypothesis is found, we can continue the process to provide a steadily growing set of estimation hypotheses that is ordered with respect to the likelihood b_k. The process can be terminated whenever the *computation time* or the *memory space* exceeds a certain limit. This so-called *any-time/any-space* formulation of the estimation algorithm guarantees that the set of hypotheses found so far represents the leading set of estimates for the specific time-step under consideration. The following sections demonstrate our novel approach to hybrid estimation in that they

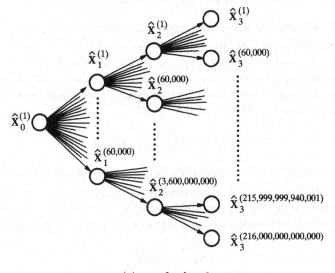

(a) tree for $k = 3$

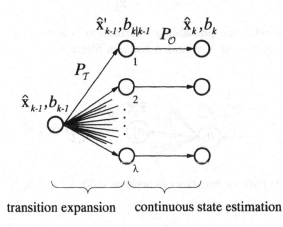

(b) node expansion

Fig. 4.8. Full hypothesis tree.

reformulate the estimation task as a search problem and provide efficient and focused solutions for it.

4.5.1 Hybrid Estimation as Shortest Path Problem

In order to reformulate hybrid estimation as search, we first take a closer look at the operations that take place whenever hybrid estimation deduces possible successors of a hybrid estimate $\hat{\mathbf{x}}_{k-1}^{(i)}$ with the likelihood $b_{k-1}^{(i)}$. Figure 4.8b visualizes the 2-step hybrid estimation process (4.71) and (4.72). The first step calculates the possible immediate successors $\hat{\mathbf{x}}_{k-1}'$, according to the transition specification of the underlying cPHA model

$$\hat{\mathbf{x}}_{k-1}^{(i)} \longrightarrow \{\hat{\mathbf{x}}_{k-1}'^{(1)}, \ldots, \hat{\mathbf{x}}_{k-1}'^{(\lambda)}\} \tag{4.98}$$

and the prior probability distribution $b_{k|k-1}$, according to (4.71). The intermediate hybrid states $\hat{\mathbf{x}}_{k-1}'^{(j)}$ represent the starting point for the estimate of the trajectory suffix for the sampling period k. The second step (estimation) considers all possible evolutions and deduces the corresponding estimates

$$\hat{\mathbf{x}}_{k-1}'^{(1)} \longrightarrow \hat{\mathbf{x}}_k^{(1)}$$
$$\vdots \tag{4.99}$$
$$\hat{\mathbf{x}}_{k-1}'^{(\lambda)} \longrightarrow \hat{\mathbf{x}}_k^{(\lambda)}$$

as well as the conditional probability distribution b_k according to (4.72). This involves the execution of λ extended Kalman filters.

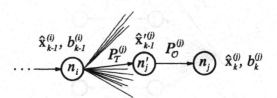

Fig. 4.9. Path for trajectory hypothesis suffix $\hat{\mathbf{x}}_{k-1}^{(i)} \to \hat{\mathbf{x}}_{k-1}'^{(j)} \to \hat{\mathbf{x}}_k^{(j)}$.

Consider, for example, the trajectory hypothesis with the fringe estimate $\hat{\mathbf{x}}_k^{(j)}$ that corresponds to a fringe node n_j in the estimation tree, as shown in Fig. 4.9. One obtains the conditional probability $b_k^{(j)}$ for the trajectory estimate that leads to $\hat{\mathbf{x}}_k^{(j)}$ by multiplying the conditional probabilities along the path and a consecutive normalization

$$b_k^{(j)} = \frac{1}{c} P_O^{(j)} P_T^{(j)} b_{k-1}^{(i)} . \tag{4.100}$$

Instead of calculating all transitions and all filters to deduce the set of hypotheses $\{\hat{\mathbf{x}}_k^{(1)}, \ldots, \hat{\mathbf{x}}_k^{(\lambda_k)}\}$, we intend to perform this operation selectively to focus on the leading set. The conditional probabilities along the possible paths in the tree indicate whether a branch in the tree leads toward a promising estimation candidate, or not. One can view this as a *shortest path* problem, where the conditional probabilities along the arcs of the tree correspond to *path lengths*, or more generally the *path costs*, and one seeks for the *shortest path*, that is, the path with the *lowest cost*. Framing the estimation problem in this context has the advantage that one can utilize advanced search algorithms that were developed for this problem, such as Dynamic Programming [16, 19] or A* [49, 50], for example.

In shortest path problems the cost of arcs along the path are combined using addition. Our probabilistic framework, however, uses multiplication (4.100) and we seek for the largest conditional probability $b_k^{(\nu)}$. Therefore, we use the the standard approach of taking the *negative logarithm* to frame the problem in the context of minimizing an additive path cost. The negative logarithm retains the ordering

$$b_k^{(\nu)} > b_k^{(j)} \iff -\ln(b_k^{(\nu)}) < -\ln(b_k^{(j)}) , \qquad (4.101)$$

as well as it transforms the multiplications to additions

$$\ln(b_k^{(j)}) = -\ln\left(\tfrac{1}{c} P_{\mathcal{O}}^{(j)} P_T^{(j)} b_{k-1}^{(i)}\right)$$

$$= \left[-\ln(P_{\mathcal{O}}^{(j)})\right] + \left[-\ln(P_T^{(j)})\right] + \left[-\ln(b_{k-1}^{(i)})\right] + \ln(c) . \qquad (4.102)$$

The normalization factor

$$c = \sum_{j=1}^{\lambda_k} P_{\mathcal{O}}^{(j)} P_T^{(j)} b_{k-1}^{(i)}$$

is the same for all trajectory hypotheses, thus it only adds a constant offset in (4.102) and can be omitted. Furthermore, it makes sense to ensure that the cost of arcs within the tree is non-negative. This ensures an monotonically increasing path cost along a possible path and simplifies the search operation. Arcs that represent transition expansions obey this property as $P_T^{(j)}$ of an admissible transition represents a (conditional) probability so that

$$P_T^{(j)} \in (0 \ \ 1] \iff -\ln(P_T^{(j)}) \in [0 \ \ \infty) . \qquad (4.103)$$

This motivates the following cost definition for transition arcs:

$$g(\text{transition-arc}_j) := -\ln(P_T^{(j)}) . \qquad (4.104)$$

A positive path cost, however, cannot be guaranteed for arcs that represent the filtering step of hybrid estimation whenever we simply take the negative logarithm of $P_{\mathcal{O}}$. Equation (4.53) defines the probabilistic observation function $P_{\mathcal{O}}$ as the value of a multi-variate Gaussian PDF

$$P_{\mathcal{O},k}^{(j)} = \frac{1}{|2\pi\mathbf{S}_k|^{1/2}} \, e^{-0.5\left[\mathbf{y}_{c,k}-\mathbf{g}^{(j)}\left(\hat{\mathbf{x}}_{c,k|k-1}^{(j)},\mathbf{u}_{c,k}\right)\right]^T \mathbf{S}_k^{-1}\left[\mathbf{y}_{c,k}-\mathbf{g}^{(j)}\left(\hat{\mathbf{x}}_{c,k|k-1}^{(j)},\mathbf{u}_{c,k}\right)\right]}\,.$$

$$(4.105)$$

This implies that $P_{\mathcal{O}}$ is not limited to positive values smaller than 1. As a consequence, we would obtain a negative path cost whenever $P_{\mathcal{O}} > 1$. Independently of this difficulty, Maybeck and Stevens [75] reported in the context of MM-estimation that the normalization term in (4.105) represents an artificial bias that can lead to incorrect mode estimates. They suggest to omit the normalization and utilize a modified observation function

$$\bar{P}_{\mathcal{O},k}^{(j)} := e^{-0.5\left[\mathbf{y}_{c,k}-\mathbf{g}^{(j)}\left(\hat{\mathbf{x}}_{c,k|k-1}^{(j)},\mathbf{u}_{c,k}\right)\right]^T \mathbf{S}_k^{-1}\left[\mathbf{y}_{c,k}-\mathbf{g}^{(j)}\left(\hat{\mathbf{x}}_{c,k|k-1}^{(j)},\mathbf{u}_{c,k}\right)\right]} \quad (4.106)$$

instead. This solution is legitimate since the second step of hybrid estimation (4.72) ensures the normalization of b_k, anyhow. The replacement of $P_{\mathcal{O}}$ with $\bar{P}_{\mathcal{O}}$ solves our cost issue as well, because the value of modified observation function $\bar{P}_{\mathcal{O}}^{(j)}$ is within the interval $[0\ 1]$. This leads to the following cost-definition for arcs that represent the filtering step within the full hypothesis tree[3]:

$$g(\text{estimation-arc}_j) := -\ln(\bar{P}_{\mathcal{T}}^{(j)})\,. \quad (4.107)$$

Up to now, we made the assumption that estimation starts from a single initial estimate $\hat{\mathbf{x}}_0$ (e.g. Fig. 4.8a). Our cPHA model, however, provides a more general initial state definition X_0 that encodes several initial estimates $\hat{\mathbf{x}}_0^{(1)}, \dots, \hat{\mathbf{x}}_0^{(\lambda_0)}$, with varying likelihood $b_0^{(1)}, \dots, b_0^{(\lambda_0)}$. As a consequence, hybrid estimation has to consider $\lambda_0 \geq 1$ hypothesis trees concurrently, each one originating from one initial estimate $\hat{\mathbf{x}}_0^{(\nu)} \in \{\hat{\mathbf{x}}_0^{(1)}, \dots, \hat{\mathbf{x}}_0^{(\lambda_0)}\}$ (Fig. 4.10a). We can interpret this situation as having *one* hypotheses tree that originates from a common (virtual) node n_0, as shown in Fig. 4.10b. An initial path segment from n_0 to a node n_ν that encodes an initial estimate $\hat{\mathbf{x}}_0^{(\nu)}$ is then characterized by the likelihood $b_0^{(\nu)}$ of the initial state. Equivalently, we can say that a node n_ν, which abstracts the initial state $\hat{\mathbf{x}}_0^{(\nu)}$, is characterized by the initial cost

$$g(n_\nu) := -\ln\left(b_0^{(\nu)}\right)\,. \quad (4.108)$$

This specification for the initial states, together with the arc costs that were introduced above, enable us to define the overall path cost from the root node n_0 to a node n_j recursively as

$$g(n_j) := -\ln(\bar{P}_{\mathcal{O}}^{(j)}) - \ln(P_{\mathcal{T}}^{(j)}) + g(n_i)\,, \quad (4.109)$$

where $P_{\mathcal{T}}^{(j)}$ denotes the probability of the transition along the trajectory suffix under consideration and $\bar{P}_{\mathcal{O}}^{(j)}$ denotes the modified observation function

---

[3] The modified observation function $\bar{P}_{\mathcal{O}}^{(j)}$ can be provided by the underlying filtering algorithm in the same way as the original one that was given in (4.53).

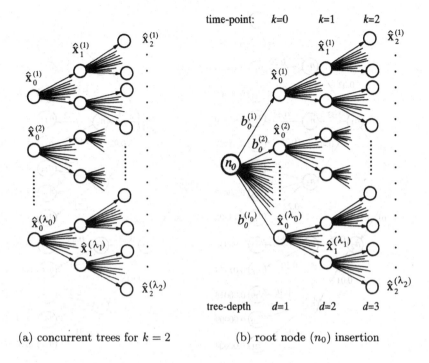

Fig. 4.10. Full hypothesis tree for several initial states $\hat{\mathbf{x}}_0^{(1)}, \ldots, \hat{\mathbf{x}}_0^{(l_0)}$.

(4.106) for the resulting estimation hypothesis j. In the following, we will also use the notation

$$g_{i,j} :- g(a_{i,j}) = g(n_j) - g(n_i) = -\ln(\bar{P}_{\mathcal{O}}^{(j)}) - \ln(P_{\mathcal{T}}^{(j)}) \qquad (4.110)$$

to denote the cost of the arc $a_{i,j} : n_i \xrightarrow{a_{i,j}} n_j$ that connects the adjacent nodes n_i and n_j. The number of arcs from the root node n_0 to a particular node n_i specifies the *tree-depth* $d(n_i)$. The tree-depth directly relates to the time-step k of a hybrid estimate $\hat{\mathbf{x}}_k^{(i)}$ that is encoded by a node n_i, namely $d = k+1$ (see Fig. 4.10b). We express the correspondence between a hybrid state estimate $\hat{\mathbf{x}}_k^{(i)}$ and an associated node n_i in terms of the functions

$$
\begin{aligned}
n_i &= \texttt{node}(\hat{\mathbf{x}}_k^{(i)}) \\
\hat{\mathbf{x}}_k^{(i)} &= \texttt{hybrid-state}(n_i) \\
k &= \texttt{time-point}(n_i) \, .
\end{aligned}
\qquad (4.111)
$$

However, whenever it is clear from the context, we sometimes use the estimate $\hat{\mathbf{x}}_k^{(i)}$ to refer to the corresponding node n_i within the hypothesis tree directly, without explicitly emphasizing the underlying mapping. Figure 4.11 illustrates

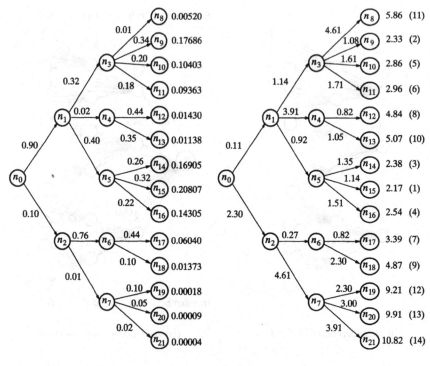

(a) tree with probability labeling

(b) tree with cost labeling and ranking

Fig. 4.11. Simple two-step full hypothesis tree with two initial states $(\{\hat{\mathbf{x}}_0^{(1)}, \hat{\mathbf{x}}_0^{(2)}\} \hat{=} \{n_1, n_2\})$ and estimates for $k = 1$ $(\{\hat{\mathbf{x}}_1^{(1)}, \dots, \hat{\mathbf{x}}_1^{(5)}\} \hat{=} \{n_3, \dots, n_7\})$ and $k = 2$ $(\{\hat{\mathbf{x}}_2^{(1)}, \dots, \hat{\mathbf{x}}_2^{(14)}\} \hat{=} \{n_8, \dots, n_{21}\})$.

the relationship between the conditional probabilities of hybrid estimation (Fig. 4.11a) and the cost of the paths (Fig. 4.11b) for a full hypothesis tree.

The recursive application of the hybrid estimation (4.71)-(4.72) can be seen as performing a *breadth-first search* on the full hypothesis tree. This search algorithm exhaustively considers all nodes at tree-depth j of the hypothesis tree, before going to the next level $(j + 1)$.

A Dynamic Programming View of Hybrid Estimation

Probably the best known algorithmic approach for solving shortest path (or minimum cost) problems is Dynamic Programming [16, 19]. Dynamic Programming (DP) relies upon Bellmans *principle of optimality* that states:

> **Principle of Optimality:** Let $\{n_0, n_1^*, n_2^*, \dots, n_k^*\}$ be the optimal (minimum cost) path from the root node n_0 to a goal node n_k^*. Then,

the truncated policy $\{n_i^*, n_{i+1}^*, \ldots, n_k^*\}$ is optimal for the subproblem that starts at n_i^*.

This rather apparent fact suggests to construct the optimal path, starting with the tail problem that considers 1-step paths to goal nodes, and then extending the paths incrementally for two, three, four, etc., steps until the path reaches the root node n_0. In terms of our application this would mean that the algorithm proceeds backward in time, which is impractical for on-line estimation. However, it is evident that the shortest path from the root node n_0 to a goal node n_k^* is also optimal for the reversed shortest path problem, which starts at the node n_k^*, traverses the arcs in the opposite direction, and that terminates at the node n_0. Interpreting a search problem in the reversed direction leads to the *forward DP* algorithm.

Forward DP starts at the root node n_0 and considers all one-step paths first (Fig. 4.12a). This operation is equivalent to ranking the initial states $\hat{\mathbf{x}}_0^{(1)}, \ldots, \hat{\mathbf{x}}_0^{(\lambda_0)}$. It then incrementally extends the paths for two, three, etc., steps until it reaches nodes n_i that abstract hybrid estimates $\hat{\mathbf{x}}_k^{(i)}$ for the time-step k under consideration (Fig. 4.12c for $k = 2$).

The advantage of DP over a *breadth-first search* technique is due to the DP's principle of optimality. Whenever several paths within a graph lead to a node n_i, DP only extends the path with the lowest cost and discards redundant paths of higher cost. The full hypothesis tree of hybrid estimation, however, does not share nodes among the branches, because each hybrid estimate $\hat{\mathbf{x}}_j^{(i)}$ belongs to one trajectory hypothesis only. This property directly follows from the fact that the state of a hybrid model can take on infinitely many valuations due to the real valued continuous state variable. As a consequence, there is only one path through each node n_i and we do not gain any advantage of using DP instead of performing breadth-first search, or directly applying the recursive hybrid estimation equations (4.71)-(4.72).

This is in contrast to discrete estimation with hidden Markov Models (HMM). Whenever one is interested in estimating the possible mode sequence for the HMM model with a finite number of l possible states (modes), one can collapse the hypothesis tree into a *Trellis diagram*. A trellis diagram is an acyclic graph that has at each layer at most l nodes, one for each mode of the model. The connecting arcs of two layers are determined by the transition topology of the model. The arcs are labeled with the product of conditional transition and the conditional observation probability, given the mode and the imperfect (discretely valued) observation at the time-point under investigation. Figure 4.13 illustrates the mapping of the hypothesis tree to the corresponding trellis diagram for a simple HMM model with 4 possible modes. For example, the nodes $\{n_8, n_{13}, n_{16}, n_{18}, n_{21}\}$ of the full hypothesis tree (Fig. 4.13a) that describe fringe states for trajectory estimates with mode m1 at $k = 2$ are mapped to a single node n_6 of the associated trellis diagram (Fig. 4.13b).

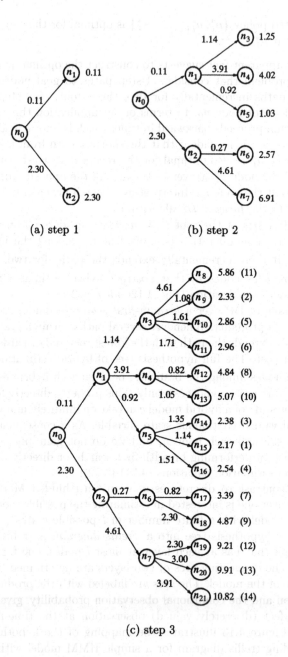

(a) step 1   (b) step 2

(c) step 3

Fig. 4.12. Expansion of full hypothesis tree with forward Dynamic Programming.

The advantage of the trellis diagram is that it does not grow beyond the bound of l nodes per layer (or estimation step). Furthermore, one can take

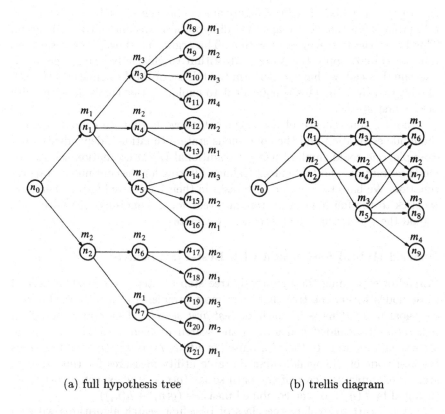

(a) full hypothesis tree   (b) trellis diagram

Fig. 4.13. 2-step full hypothesis tree for HMM and associated trellis diagram.

advantage of the principle of optimality and only consider the most likely path
from the source n_0 to a node n_i within the tree. This forward DP approach for
identifying the most likely mode sequence is known as the *Viterbi* algorithm in
literature [19]. Even though it is tempting to apply this approach for hybrid
estimation as well, for example, to identify the most likely mode sequence
only, it is easy to see that this cannot be done. The hybrid model captures
mode transitions and observations that are dependent of the continuous state.
Thus, P_O and P_T and the corresponding cost $g(a)$ of the associated arc within
the tree, depend on $\mathbf{x}_c$. Even though two paths of the full hypothesis tree lead
to two estimates $\hat{\mathbf{x}}_\nu^{(i)}$ and $\hat{\mathbf{x}}_\nu^{(j)}$ with the same mode $\mathbf{m}_\iota$, they will almost
always have different continuous estimates $\hat{\mathbf{x}}_{c,\nu}^{(i)} \neq \hat{\mathbf{x}}_{c,\nu}^{(j)}$. Now let us assume
that these distinct hybrid states are abstracted as a single node n_ι in a trellis
diagram because of the same mode. Their different continuous state estimates
imply different costs for the continuation $n_\iota \rightarrow \ldots \rightarrow n_{\text{goal}}$. This violates the
principle of optimality, because optimality of the trajectory suffix depends on

the trajectory prefix[4]. The IMM algorithm can be seen as a sub-optimal hybrid estimation algorithm that maps the full hypothesis tree into a trellis diagram. This limits the tree growth over time. The argument above, however, shows that we cannot apply the Viterbi algorithm to efficiently search the trellis diagram. Instead, we have to perform the full set of IMM calculations ((4.89)-(4.95)) step by step. This is equivalent to exploring the Trellis diagram with breath-first search.

Another consequence of the argument above is that we cannot perform the mode estimation and the continuous state estimation independently. For example, one could think of using a traditional HMM estimation scheme for mode estimation, and (extended) Kalman filtering for the continuous state estimation. We have to consider both tasks in their interleaved form. Therefore, we stick to the full hypothesis tree and seek for mechanisms that selectively extend the tree toward the most likely estimates.

Focused Hybrid Estimation with Best-First Search

Instead of expanding the hypothesis tree step by step, we intend to expand those nodes within the tree that seem most promising to lead toward a low-cost goal node. This is in spirit of *best-first search*[5]. This search paradigm maintains unexpanded nodes n_i of the search problem in a search agenda, sorts them according to their 'promise', or *utility measure* $f(n_i)$, and expands the *best* node of the agenda *first*. Suitable utility measures for this strategy are, for example, the cost of the path so far $(g(n_i))$, the expected cost to go (denoted by $h(n_i)$), or the combined measure $(g(n_i) + h(n_i))$.

Let us start to explore the class of best-first search algorithms with an algorithm that utilizes the cost $f(n_i) := g(n_i)$ of the path so far to select the node that is expanded in the next search step. The algorithm can be seen as a variant of the single-source shortest path algorithm of *Dijkstra* [34, 3] and is also referred to as *uniform-cost* search [89] in literature.

Again, we use the simple two-step full hypothesis tree of Fig. 4.11 as the example to illustrate the search algorithm. The initialization of the best-first search problem puts the root node n_0 onto the search agenda. The first search step removes n_0 from the agenda and deduces all possible successors. These successors are then put back onto the agenda, sorted by their utility value. This operation is equivalent to deducing all possible initial states $\hat{x}_0^{(i)}$ and sorting them according to their belief $b_0^{(i)}$. In our example this leads to (Fig. 4.14a)

---

[4] This is directly related to the fact that the consideration of the mode only represents a coarse quantization of the hybrid state that implies the loss of the Markov property [70].

[5] Some literature sources use the term *best-first* search to denote a specific algorithm that utilizes the expected cost to go as utility (e.g. [108]). We adopt the notation of [89] and denote a *class* of algorithms that utilize a variety of definitions for the utility of a node.

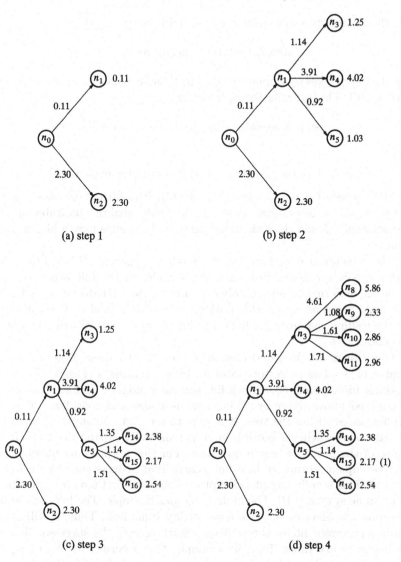

Fig. 4.14. Expansion of full hypothesis tree with uniform-cost search until the first solution is found.

$$\text{step 1: agenda} = \{n_1, n_2\}\,.$$

The second step, takes the leading unexpanded node n_1 from the agenda and applies the goal test, in other words, it evaluates whether n_1 represents an estimate at the time-step under consideration (e.g. $k = 2$). The goal test fails, and search proceeds by determining the successor nodes of n_1. This expansion returns $\{n_3, n_4, n_5\}$ (Fig. 4.14b). Search inserts these nodes in the agenda according to their utility values and returns

$$\text{step 2: agenda} = \{n_5, n_3, n_2, n_4\} \; .$$

The next step takes n_5, applies the goal test, and expands n_5 to $\{n_{14}, n_{15}, n_{16}\}$ (Fig. 4.14c). This provides the updated agenda

$$\text{step 3: agenda} = \{n_3, n_{15}, n_2, n_{14}, n_{16}, n_4\} \; ,$$

and

$$\text{step 4: agenda} = \{n_{15}, n_2, n_9, n_{14}, n_{16}, n_{10}, n_{11}, n_4, n_8\}$$

for the consecutive search step (Fig. 4.14d). The fifth step takes n_{15} and recognizes the node as a goal node. The successful goal test indicates that the first solution is found since all other nodes on the agenda have a higher utility value.

The advantage over breath-first search is apparent. While breath-first search expands 7 nodes to obtain all 21 nodes of the full hypothesis tree for $k = 2$, uniform-cost search only expands 4 nodes and identifies the leading goal node based on a partial hypothesis tree with 12 nodes. Thus, it identifies the leading goal-node (or hybrid estimate) by only considering half of the hypothesis tree.

One reason for the rapid growth of the full hypothesis tree is the large number of possible successor states for hybrid estimates (Fig. 4.8 illustrates this issue for a system 60,000 possible successor modes). Uniform-cost search, as outlined above, however, fully expands nodes and considers all successors for nodes within the tree. It is easy to see that one ought to avoid this full expansion whenever possible, and incrementally expand the nodes in the course of the best-first search operation. For this purpose, we utilize a progressive implementation of best-first search that deduces successor nodes on demand. This search variant is in spirit of the best-first search engine in the Livingstone system [104]. The underlying idea is simple. The best-first search operation considers nodes with lower utility value first. Thus, it will always consider successor nodes consecutively, starting with the successor that has the lowest utility value. Take, for example, the expansion $n_1 \rightarrow \{n_3, n_4, n_5\}$ that is shown in Fig. 4.14b. The leading successor node n_5 (utility 1.03) will always be considered prior the other successors n_3 and n_4 (utility 1.25 and 4.02, respectively). As a consequence, we can delay the deduction of n_3 until the node n_5 is selected for further expansion. Whenever this happens, we perform a *follow-up expansion* of the *predecessor node* n_1, together with the first expansion of n_5. This operation provides n_3 and n_{15} and puts them onto the agenda. The operation of this *progressive* best-first search algorithm is illustrated in Fig. 4.15.

The first step expands the root node n_0. This provides the leading successor n_1

$$\text{step 1: agenda} = \{n_1\} \; .$$

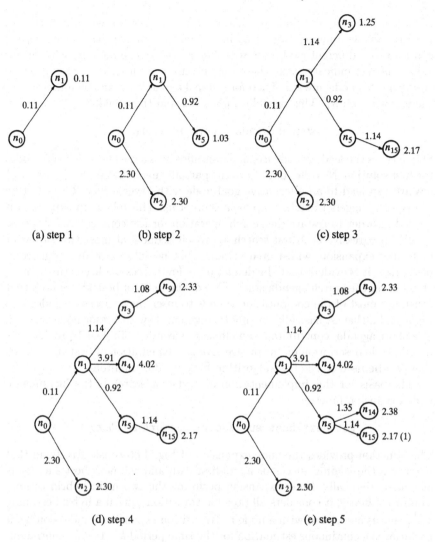

Fig. 4.15. Expansion of hypothesis tree with progressive uniform-cost search until the first solution is found.

Since n_1 is the only node on the agenda it is selected for further expansion, triggering a follow-up expansion of n_0. This yields the successor nodes n_5 and n_2, respectively.

$$\text{step 2: agenda} = \{n_5, n_2\} .$$

The third step expands n_5. Search obtains n_{15} for the expansion and n_3 for the follow-up expansion of n_1.

$$\text{step 3: agenda} = \{n_3, n_{15}, n_2\}$$

Although n_{15} is a goal node (even the leading one as we will see later), we need the follow up expansion of n_1 in order to guarantee that no other path can lead to a different goal node with lower utility. The node n_3 with utility $1.25 < g(n_{15})$ indicates that there might exist a path with a possibly lower cost from this node onward. Therefore, step 4 expands n_3, and again, performs the follow-up on n_1. This provides n_9 and n_4 and the agenda

$$\text{step 4: agenda} = \{n_{15}, n_2, n_9, n_4\} \ .$$

Step 5 takes the leading node n_{15} and identifies it as a goal node. This provides the first solution. No other path from the partially expanded nodes $\{n_3, n_4, n_5\}$ onward can lead to an alternative goal node with lower utility. The selection of n_{15} also triggers the follow-up expansion of n_5. This follow-up expansion is needed in order to restart the search operation for the consecutive solutions.

The progressive best-first search approach deduces at most two successor nodes per expansion, whilst guaranteeing that no solution is overseen (completeness). It is evident that the deduction of few successors keeps the number of nodes on the search agenda small. One of the tasks of best-first search that consumes most of the computation time is to insert new successor nodes into the agenda, that is, basically sorting the agenda. Few successor nodes, as well as a short agenda, contribute to an efficient execution. Table 4.1 provides the pseudo-code description of our progressive implementation of best-first search that will be used as the base-algorithm for hybrid estimation.

The basis for this implementation of best-first search is the incremental node expansion function

get-next-best-successor (n_i, *search-problem*) .

This function provides the node expansion of Fig. 4.8b consecutively, in that it returns the leading successor at its first call, and the next best successors at consecutive calls. Node expansion performs the two main hybrid estimation tasks: firstly, it considers all possible transitions, given a hybrid estimate $\hat{\mathbf{x}}_{k-1}^{(i)}$ that is associated with a node n_i (transition expansion), and secondly, it performs the continuous estimation for the time period $k-1 \rightarrow k$ (continuous state filtering). Branching occurs at the transition expansion step. The potentially large set of possible transitions is due to the combinatorial combination of few transitions for each component (e.g. 10 components with 3 possible transition each leads to $3^{10} \approx 60,000$ possible successors, as cited above). In our cPHA modeling framework, we assume that the mode transitions of the system's components are independent of each other. This assumption allows us to consider possible mode transitions component-wise and condition them on the hybrid estimate $\hat{\mathbf{x}}_{k-1}^{(i)}$ and the actuated input $\mathbf{u}_{k-1}$. Component-based approaches for identifying mode transitions was introduced within the Livingstone [104] system, based on algorithms developed within the GDE [29] and Sherlock [30] systems for multiple fault diagnosis. We pursue a similar approach for hybrid estimation/diagnosis and formulate the enumeration of

Table 4.1. Pseudo-code for progressive best-first search.

---

function progressive-best-first-search(*problem*)
  returns the next best solution and updated search problem or signals failure
  solution ← { }
  agenda ← agenda[*problem*]
  $f(\cdot)$ ← utility-function[*problem*]
  while *solution* = { }
    if *agenda* = { }
    return { { }, *problem*}
    else
      node ← remove-best-node(*agenda*)
      follow-up-node ← get-next-best-successor(predecessor(*node*), *problem*)
      agenda ← insert-node(*follow-up-node*, *agenda*, $f(\cdot)$)
      if goal-test(*node*) = **True**
        solution ← *node*
      else
        successor-node ← get-next-best-successor(*node*, *problem*)
        agenda ← insert-node(*successor-node*, *agenda*, $f(\cdot)$)
  solutions[*problem*] ← append(solutions[*problem*], *solution*)
  agenda[*problem*] ← *agenda*
  return {*solution*, *problem*}

---

the possible successors as an underlying search problem Fig. 4.16 illustrates
this operation for a cPHA with ζ components.

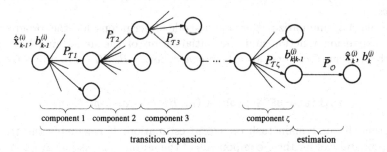

Fig. 4.16. Best-first search tree for component-wise node expansion (with probability labeling).

Starting with the hybrid estimate $\hat{\mathbf{x}}_{k-1}^{(i)}$, and the actuated input $\mathbf{u}_{k-1}$,
we determine the set of possible transitions and their associated transition
probabilities $P_{\mathcal{T}i}$, for every component $\mathcal{A}_i$ of the cPHA independently. These
transitions define the first ζ layers of the associated search tree (Fig. 4.16).

Node expansions at the last layer $(\zeta + 1)$ of the tree performs the continuous estimation and provides the updated hybrid estimates $\hat{\mathbf{x}}_k^{(j)}$. Let us illustrate this operation with a simple cPHA. The cPHA shall be comprised of 3 components $\mathcal{A}_1, \mathcal{A}_2, \mathcal{A}_3$, where each component can be in one out of two modes. So any hybrid estimate $\hat{\mathbf{x}}_{k-1}^{(i)}$ can have at most $2^3 = 8$ successors $\hat{\mathbf{x}}_k^{(j)}$. We assume that the estimate $\hat{\mathbf{x}}_{k-1}^{(i)}$ denotes a hybrid state with mode $\hat{\mathbf{x}}_{d,k-1}^{(i)} = [m_{11}, m_{21}, m_{31}]^T$ and consider the following transitions

$$
\begin{aligned}
\mathcal{A}_1 : m_{11} &\rightarrow m_{11} : P_{T11} = 0.70 \\
m_{11} &\rightarrow m_{12} : P_{T12} = 0.30
\end{aligned}
$$

$$
\begin{aligned}
\mathcal{A}_2 : m_{21} &\rightarrow m_{21} : P_{T21} = 0.55 \\
m_{21} &\rightarrow m_{22} : P_{T22} = 0.45
\end{aligned} \tag{4.112}
$$

$$
\begin{aligned}
\mathcal{A}_3 : m_{31} &\rightarrow m_{31} : P_{T31} = 0.60 \\
m_{31} &\rightarrow m_{32} : P_{T32} = 0.40 .
\end{aligned}
$$

These transitions and the consecutive filtering step spans a tree as shown in Fig. 4.17 (with hypothesized $\bar{P}_O$ values for the filtering steps). Again, we do apply best-first search to enumerate the possible successors $\hat{\mathbf{x}}_k^{(j)}$ of $\hat{\mathbf{x}}_{k-1}^{(i)}$, or the associated search nodes of the main search problem, respectively. The function get-next-best-successor$(\cdot)$ of the main search problem implements the progressive successor enumeration. The function initiates the underlying best-first search problem and searches for the first solution, upon its first call. It returns this solution, which represents the least-cost successor of n_i, and stores the search agenda for consecutive successor generations. Each subsequent call to the node expansion function restarts the search and provides the next best successor.

Table 4.2 summarizes the pseudo-code descriptions for the progressive node expansion function and the initialization of the underlying best-first search problem (*best-first successor generation – BFSG*). The initialization function

initiate-BFSG-problem$(n_i, BFSG$-*search-problem*$)$

performs the first main task of hybrid estimation. It calculates the transition probabilities for the ζ components of the automaton, as indicated in Section 4.3, and sets up the search problem accordingly. This underlying best-first search problem utilizes an instance of the progressive node expansion function progressive-best-first-search$(\cdot)$ for nodes of *class BFSG-Node*. It builds a search tree that is based on the component transitions up to a tree-depth of ζ. The node expansion at the layer ζ performs the underlying continuous estimation/filtering. This involves the deduction of a suitable filter, according to the mode hypothesis and the cPHA model (or its retrieval from a database), as well as the filter execution that provides the new estimate $\hat{\mathbf{x}}_k^{(j)}$. A

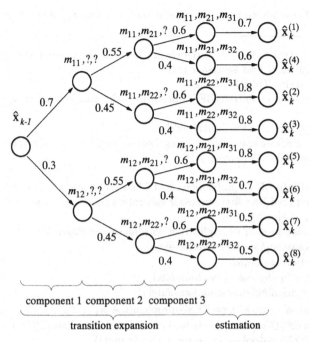

Fig. 4.17. Search tree for successor node enumeration with mode and probability labeling.

slightly different treatment is necessary whenever we expand the *root* node of the tree. This expansion performs a best-first search problem that utilizes the initial state information $X_{0\nu}$ of the individual automaton components $\mathcal{A}_\nu$ and combines them to deduce a node that represents the next best initial state $\hat{\mathbf{x}}_0^{(j)}$.

Up to now we performed uniform-cost search in that we utilize the cost from the root of the tree to a node $(g(n_\nu))$ as the utility function for best-first search. Completeness of uniform-cost search grounds onto the fact that a valid solution of the search problem has superior utility, compared to all other nodes on the agenda, regardless of their positions within the search tree. Since the utility increases monotonically along a search path, it is easy to see that nodes in the vicinity of the root node tend to have low utility, thus delay the solution labeling for terminal nodes of the tree. Take, for example, the node expansion introduced above. The leading successor with mode $[m_{11}, m_{21}, m_{31}]^T$ is deduced after 9 steps, however, it takes additional 5 steps to ensure that this successor is the first solution (Fig. 4.18).

We can focus the search even more whenever we can estimate the cost for the best path from an intermediate node n_ν to a goal node. Of course, this should be done heuristically, without performing the costly filtering operation. Given the transition probabilities for the individual components (4.112), it

Table 4.2. Pseudo-code for progressive node expansion of hybrid estimation.

---

function get-next-best-successor(*node* of Class *HE-Node, main-search-problem*)
    returns the next best successor node
    if first-expansion?(*node*) = True
        BFSG-problem ← initiate-BFSG-problem(*node, main-search-problem*)
    else
        BFSG-problem ← BFSG-data[*node*]
    {*successor*, BFSG-data[*node*]} ← progressive-best-first-search(*BFSG-problem*)
    return new-HE-Node(*successor*)

function initiate-BFSG-problem(*node* of Class *HE-Node, main-search-problem*)
    returns initiated best-first successor generation search problem
    k ← time-point[*node*]
    hybrid-estimation-data ← application-data[*main-search-problem*]
    $\mathcal{CA}$ ← automaton[*hybrid-estimation-data*]
    $\hat{\mathbf{x}}_{k-1}^{(i)}$ ← hybrid-state[*node*]
    U ← input-value[*hybrid-estimation-data*]
    Y_c ← observation[*hybrid-estimation-data*]
    BFSG-problem ← make-new-BFSG-problem($\mathcal{CA}, \hat{\mathbf{x}}_{k-1}^{(i)}, U_{k-1:k}, Y_{c,k-1:k}$)
    transitions[*BFSG-problem*] ← deduce-component-transitions($\hat{\mathbf{x}}_{k-1}^{(i)}, U_{k-1:k-1}, \mathcal{CA}$)
    agenda[*BFSG-problem*] ← {new-BFSG-Node(root)}
    return *BFSG-problem*

---

is relatively easy to deduce the best possible transition for each of them. Whenever we are interested in the expected cost to go, we can utilize the best transitions for the remaining components to obtain a conservative estimate for the transition expansions. We cannot predict the outcome of the filtering step without performing the filtering operation. Therefore, we assume the upper conservative bound $\bar{P}_O = 1$ for the estimation step. This value, however, does not provide any useful information as $\ln(\bar{P}_{O\max}) = -\ln(1) = 0$. Therefore, we draw the value of the expected cost to go from the transition steps only.

Consider the 3-PHA example with the possible transitions listed in (4.112). Given the partial mode assignment $[m_{12}, ?, ?]^T$ of the node n_2, the best possible completion is to select the transitions $m_{21} \to m_{21}$ and $m_{31} \to m_{31}$ for the remaining PHA components $\mathcal{A}_2$ and $\mathcal{A}_3$, respectively. This leads to an expected cost to go of

$$h(n_2) = -\ln(P_{T21}) - \ln(P_{T31}) = 1.11 \ . \tag{4.113}$$

More generally, we can formulate the heuristic $h(\cdot)$ for the expected cost to go from a node n to a goal node as

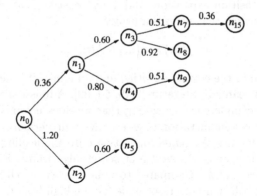

(a) goal-node deduction

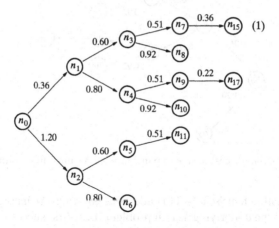

(b) solution deduction

Fig. 4.18. Partial successor tree expansion until first solution is found.

$$h(n) := \sum_{i=\iota+1}^{\varsigma} -\ln(\max_{\eta} P_{Ti\eta}) \, , \qquad (4.114)$$

where $\iota = d(n)$ denotes the number of components with assigned mode, and ς denotes the number of components of the cPHA model. The heuristic (4.114) is an *admissible heuristic*, that is, it *never overestimates* the cost to reach the goal node, since it considers the best possible transitions that imply the lowest cost for each transition expansion, and a zero cost estimate for the filtering step. Best-first search that utilizes a utility

$$f(n_i) = g(n_i) + h(n_i) , \qquad (4.115)$$

that is composed of the cost so far $g(n_i)$ and an admissible heuristic $h(n_i)$, is referred to as A* search in literature[6] [49, 50, 89]. Admissibility of the heuristic $h(\cdot)$ ensures completeness, meaning that A* does not miss solutions and provides the correct enumeration of goal nodes. Furthermore, the application of the heuristic focuses the search operation onto the leading solutions, thus improves the efficiency of the best-first search algorithm[7]. Figure 4.19 illustrates the operation of A*. Compared to uniform-cost search (Fig. 4.18b), A* deduces the first solution (successor node for the high-level hybrid estimation search problem) by considering fewer nodes.

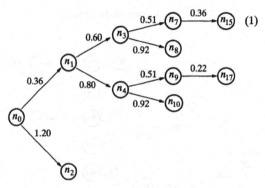

Fig. 4.19. Partial successor tree expansion with A* until first solution is found.

The admissible heuristic (4.114) allows us to utilize A* instead of uniform-cost search for the underlying search problem (best-first successor generation). Similarly, we also would like to use A* for the main search problem that implements focused hybrid estimation. For this purpose, we have to obtain an admissible heuristic for the main hybrid estimation problem. Transition deduction, in particular guard probability calculation, and the continuous estimation (filter deduction and execution) are the two tasks of hybrid estimation that consume most of the computation time. Therefore, it is evident that a suitable admissible heuristic should avoid these operations in the course of

---

[6] To be precise, A* denotes a best-first search algorithm that utilizes the utility (4.115), as well as it applies the *dynamic programming principle*, to eliminate redundant paths with higher cost. The latter does not provide any advantages for our hybrid estimation application and, therefore, is not considered here.

[7] In fact, one can show that A* is complete, optimal and optimally efficient for any given admissible heuristic function. See, for example, [89] for an intuitive proof for these properties.

obtaining a lower upper-bound for expected the transition probabilities and values for the observation function. As for (4.114) we have to take the default value $\bar{P}_{\mathcal{O}} = 1$ for the observation function. For the transition probabilities, however, it is possible to obtain lower bounds on them without evaluating the guards. For this purpose, we have to reconsider the transition specifications of our PHA models (Def. 3.2). A transition for a cPHA component is defined in terms of the set-valued function $T : \mathcal{X}_d \rightarrow 2^{\mathcal{T}}$ that assigns a set of possible transitions for each mode $m_i \in \mathcal{X}_d$. These transitions are specified in terms of *transition triples* $\tau_i := \langle p_{\tau i}, c_i, r_i \rangle \in \mathcal{T}$. Its probability mass function $p_{\tau i}$ defines the *threads* of the transition τ_i that is guarded with the conditional guard c_i. The threads define the goal modes, as well as the conditional transition probabilities, given c_i is satisfied. Let us reconsider the visualization of this transition concept in Fig. 4.20 (previously given in Fig. 3.2). The function T specifies the two transitions $T(m_1) = \{\tau_1, \tau_2\}$ for mode m_1. Transition τ_1, for example, has the associated guard c_1 and a probability mass function $p_{\tau 1}$, that encodes the conditional probabilities P_1 and P_2 of Fig. 4.20. Without knowing the truth values (or conditional probabilities) for the two guards, we can conclude that any transition probability $P_{\mathcal{T}i}$ for transitions out of mode m_1 is limited by

$$P_{\mathcal{T}i} \leq \max\{P_1, P_2, P_3, P_4\} =: P_{\mathcal{T}\max}(m_i) . \qquad (4.116)$$

In words, the thread with the largest conditional probability specifies a bound for the actual transition probabilities. Let us assume that we consider a node n_i of the full hypothesis tree that represents a hybrid estimate $\hat{\mathbf{x}}_{k-j}^{(i)}$ for the time-step $k - j$. Based on the mode $\hat{\mathbf{x}}_{d,k-j}^{(i)}$ of the estimate $\hat{\mathbf{x}}_{k-j}^{(i)}$, we can infer an upper bound for the transition probability of

$$P_{\mathcal{T}\max}(\hat{\mathbf{x}}_{d,k-j}^{(i)}) = \prod_{\nu=1}^{\zeta} P_{\mathcal{T}\max}(\hat{\mathbf{x}}_{d\nu,k-j}^{(i)}) , \qquad (4.117)$$

where $\hat{\mathbf{x}}_{d\nu,k-j}^{(i)}$ denotes the mode estimate for the cPHA automaton component $\mathcal{A}_\nu$ at $k - j$.

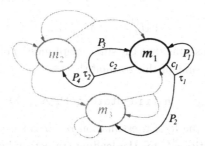

Fig. 4.20. Guarded probabilistic mode transitions of a PHA.

No mode estimate is available for the transitions beyond the time-step $k - j$. Therefore, we can only base the estimate on all possible transitions of each component. The value of the most likely thread of a component $\mathcal{A}_\nu$ specifies the lower upper bound $P_{T\max}(\mathcal{A}_\nu)$ for the component's transitions and provides the value

$$P_{T\max}(\mathcal{CA}) = \prod_{\nu=1}^{l} P_{T\max}(\mathcal{A}_\nu) \tag{4.118}$$

for the cPHA automaton.

Given a node n_i with its associated hybrid estimate $\hat{\mathbf{x}}_{k-j}^{(i)}$ that specifies the mode $\hat{\mathbf{x}}_{d,k-j}^{(i)}$, we define the following admissible heuristic for the continuation up to a time-step k:

$$h(n_i) := -\ln(P_{T\max}(\hat{\mathbf{x}}_{d,k-j}^{(i)})) - (j-1)\ln(P_{T\max}(\mathcal{CA})) . \tag{4.119}$$

The leading threads for the PHA components, as well as for the PHA components at a particular mode, can be determined at compile-time of the cPHA model. As a consequence, the evaluation of (4.119) is computationally efficient. This enables us to utilize A* for the main search problem as well. Overall, we obtain a search procedure which highly focuses onto the portion of the full hypothesis tree that contains the leading trajectory estimates. The following table (4.3) summarizes this approach for hybrid estimation in pseudo-code. The core function hybrid-estimation($\cdot$) executes best-first search until it used up the computational resources for hybrid estimation, or the number of solutions reaches a user defined upper bound ($\eta_{k,\max}$). The search operation returns an ordered set of $\eta_k \leq \eta_{k,\max}$ search nodes

$$\mathcal{N}_k = \{n_i, \ldots, n_j\} \tag{4.120}$$

that represent the fringe estimates

$$\mathcal{X}_k := \{\hat{\mathbf{x}}_k^{(1)}, \ldots, \hat{\mathbf{x}}_k^{(\eta_k)}\} = \text{hybrid-state}(\mathcal{N}_k) \tag{4.121}$$

of the trajectory estimates $\{\hat{X}_k^{(1)}, \ldots, \hat{X}_k^{(\eta_k)}\}$ for the leading trajectory hypotheses

$$\{M_{d,k}^{(1)}, \ldots, M_{d,k}^{(\eta_k)}\} . \tag{4.122}$$

For simplicity, let us assume that the superscript index j of the hybrid estimate $\hat{\mathbf{x}}_k^{(j)}$ also denotes the *rank* of this estimate[8]:

$$\hat{\mathbf{x}}_k^{(j)} = \text{hybrid-state}(\text{n-th}(\mathcal{N}_k, j)) .$$

In order to interpret the hybrid estimation result correctly, we also need the conditional probabilities $b_k^{(j)}$ for the leading trajectory hypotheses

---

[8] The function n-th($list, j$) simply retrieves the j'th element of the ordered list.

Table 4.3. Pseudo-code for focused hybrid estimation.

---

function initiate-HE-problem($\mathcal{CA}$)
  returns initiated main search problem for hybrid estimation
  HE-problem $\leftarrow$ new-best-first-search-problem()
  time-step[application-data[*HE-problem*]] $\leftarrow$ 0
  automaton[application-data[*HE-problem*]] $\leftarrow$ $\mathcal{CA}$
  utility-function[*HE-problem*] $\leftarrow$ HE-utility-function($\mathcal{CA}$)
  agenda[*HE-problem*] $\leftarrow$ {new-HE-Node(root)}
  return *HE-problem*

function hybrid-estimation($\mathbf{u}_k$, $\mathbf{y}_{c,k}$, $\eta_{k,\max}$, *HE-problem*)
  returns the leading set of estimates for time-step k
  time-step[application-data[*HE-problem*]] $\leftarrow$ +1
  input-value[application-data[*HE-problem*]] $\leftarrow$ add-value($\mathbf{u}_k$)
  observation[application-data[*HE-problem*]] $\leftarrow$ add-value($\mathbf{y}_{c,k}$)
  agenda[*HE-problem*] $\leftarrow$
    append(solutions[*HE-problem*], agenda[*HE-problem*])
  solutions[*HE-problem*] $\leftarrow$ { }
  do progressive-best-first-search(*HE-problem*)
    until (resources-used-up?() **or**
        count(solutions[*HE-problem*]) $= \eta_{k,\max}$)
  return {hybrid-estimate(solutions[*HE-problem*]), HE-problem}

$$\{M_{d,k}^{(1)}, \dots, M_{d,k}^{(\eta_k)}\} .$$

Best-first search provides the path-cost $g(n_i)$ for nodes $n_i \in \mathcal{N}_k$. This path cost directly relates to the *un-normalized likelihoods*

$$\bar{b}_k^{(i)} := \bar{P}_{\mathcal{O}}^{(i)} P_{\mathcal{T}}^{(i)} \bar{b}_{k-1}^{(i)}, \qquad \bar{b}_0^{(i)} = b_0^{(i)} , \tag{4.123}$$

according to the logarithmic transformation

$$g(\mathbf{node}\,(\hat{\mathbf{x}}_k^{(i)})) = -\ln(\bar{b}_k^{(i)}) . \tag{4.124}$$

Hybrid estimation that is based on best-first search considers the η_k leading estimates out of the λ_k trajectory hypotheses of full-hypothesis hybrid estimation. As a consequence, we perform normalization upon the truncated set of estimates and obtain

$$b_k^{(i)} = \frac{\bar{b}_k^{(i)}}{\sum_{j=1}^{\eta_k} \bar{b}_k^{(j)}} . \tag{4.125}$$

These conditional probabilities, together with the fringe estimates, define an *approximation* for the hybrid belief state $b_k(\cdot)$. This involves the weighted combination of the fringe estimates $\mathcal{X}_k$ according to (4.74)-(4.79). For example, to obtain the most likely mode, or the most likely continuous estimate, given the most likely mode, or the overall continuous estimate for the time-step k.

4.5.2 Suboptimal Search Methods for Hybrid Estimation

We intend to apply hybrid estimation as the monitoring and diagnosis tool within a process automation system and operate over a considerably long period of time. A* does a good job in focusing the search operation onto the most likely branches of the full hypothesis tree. However, the tree fraction under consideration still grows monotonically as the time proceeds. This is impractical for on-line estimation that operates over a long period of time. Therefore, we modify our hybrid estimation algorithm and take a *sliding window* approach that only grows a (focused) hypothesis tree over a fixed period of time.

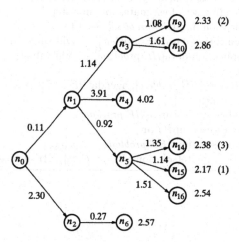

Fig. 4.21. Focused hypothesis tree for the first 3 hypotheses at $k = 2$.

N-Step Hybrid Estimation

Let us assume a fixed window-size of N steps, meaning that we grow a hypothesis tree up to a *tree-depth* of at most $N + 1$. N-step hybrid estimation grows this tree starting from a truncated set of κ leading estimates at time-step $k - N$. Discarding the less likely hypotheses at $k - N$ ensures that the tree does not grow beyond bounds, however, it also imposes *sub-optimality* as we abandon hypotheses and will not reconsider them anymore. The overall operation of N-step hybrid estimation is explained best in terms of a simple example. For this purpose, we recall the estimation example with the full hypothesis tree of Fig. 4.11, and perform N-step estimation with a window-size of $N = 2$. Let us further assume that we start the hybrid estimation for the time-step k from the $\kappa = 2$ leading hybrid state estimates at $k - N$. Figure 4.21 shows the partially grown hypothesis tree for the prediction of the $\eta_2 = 3$ leading estimates at the time-step $k = 2$ (previously given in

Fig. 4.11). Let us assume that stringent computational constraints stopped the estimation for the time-step $k = 2$ after the first three estimates were found and hybrid estimation moves on to estimate the trajectory hypothesis for $k = 3$. N-step hybrid estimation restarts by manipulating the hypothesis tree of the previous time-step $k = 2$ (Fig. 4.21). It uses the tree to determine the leading set of κ hybrid state estimates at the time-step $k - N = 1$. The associated nodes

$$\{n_5, n_3\} =: \bar{\mathcal{N}}_1 \, ,$$

of the leading estimates

$$\{\hat{\mathbf{x}}_1^{(1)}, \hat{\mathbf{x}}_1^{(2)}\}$$

serve as the 'initial nodes' for the hypothesis tree that is used to determine the hybrid state estimates at the time-step $k = 3$. Of course, we re-use tree fractions as much as possible and keep all tree branches that originate from the nodes $n_i \in \bar{\mathcal{N}}_1$. All other branches are discarded and N-step hybrid estimation will not reconsider their associated trajectory hypotheses anymore (Fig. 4.22a indicates those tree branches in gray). In the same way as we combined multiple concurrent full hypothesis trees for estimation problems with multiple initial estimates $\hat{\mathbf{x}}_0^{(i)}$, we insert a new root node to combine the κ tree fractions into a single truncated hypothesis tree (Fig. 4.22b). Since we truncate less likely estimates at the time-step $k - N$ and re-initiate the search with the κ leading estimates $\{\hat{\mathbf{x}}_{k-N}^{(1)}, \ldots, \hat{\mathbf{x}}_{k-N}^{(\kappa)}\}$ only, we have to re-normalize their conditional probabilities $b_{k-N}^{(j)}$, $j = 1, \ldots, \kappa$ as follows

$$b_k^{(i)} = \frac{\bar{b}_k^{(i)}}{\sum_{j=1}^{\kappa} \bar{b}_k^{(j)}} \, . \tag{4.126}$$

$\bar{b}_k^{(i)}$ denotes the un-normalized likelihood as given in (4.123). For example, the Fig. 4.22 specifies the following values for the unnormalized likelihoods $\bar{b}_1^{(i)}$ of $\hat{\mathbf{x}}_1^{(1)}$ and $\hat{\mathbf{x}}_1^{(2)}$

$$\bar{b}_1^{(1)} = e^{-g(n_5)} = 0.360, \qquad \bar{b}_1^{(2)} = e^{-g(n_3)} = 0.288 \, .$$

Applying (4.126), we obtain

$$b_1^{(1)} = \frac{0.360}{0.288 + 0.360} = 0.556, \qquad b_1^{(2)} = \frac{0.288}{0.288 + 0.360} = 0.444 \, .$$

These re-normalized conditional probabilities specify the costs for the arcs from the new root node n_0 to the nodes n_5 and n_3 (Fig. 4.22b):

$$g(n_5) = -\ln(0.556) = 0.59, \qquad g(n_3) = -\ln(0.444) = 0.81 \, .$$

The purpose of having a window-size $N > 1$ is to allow a low probability trajectory estimate to increase in weight, as time proceeds. Blindly truncating

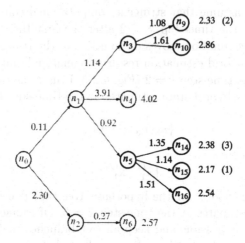

(a) tree fragments that originate from
the leading 2 estimates at $k = 1$

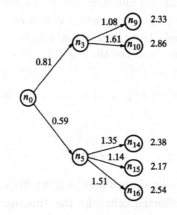

(b) sub-optimal search tree initialization for
the next estimation step ($k = 4$)

Fig. 4.22. N-Step hybrid estimation.

the tree, as outlined above, can prune fractions of the tree that represent leading trajectory estimates $\hat{X}_{k-1}^{(i)}$ for the time-step $k-1$ whenever they trace back to estimates $\hat{x}_{k-N}^{(j)}$ that are not within the set of the κ leading estimates at the time-step $k - N$. Figure 4.23 visualizes such a situation for N-step hybrid estimation with a window-size of $N = 3$ and $\kappa = 2$ (the figure utilizes a slightly different representation for the partial tree that arranges the nodes for hypotheses $\hat{x}_k^{(i)}$ vertically, in the order of their (decreasing) conditional

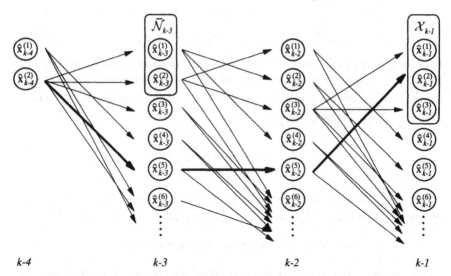

Fig. 4.23. Problematic tree truncation for N-step hybrid estimation ($N = 3, \kappa = 2$) that excludes the continuation of the estimate for the second likely trajectory hypothesis $X_{k-1}^{(2)}$.

probability $b_k^{(i)}$). The estimate $\hat{X}_{k-1}^{(2)}$ of the second likely trajectory hypothesis at $k-1$ denotes the sequence of hybrid estimates

$$\hat{X}_{k-1}^{(2)} = \{\ldots, \hat{\mathbf{x}}_{k-3}^{(5)}, \hat{\mathbf{x}}_{k-2}^{(5)}, \hat{\mathbf{x}}_{k-1}^{(2)}\} \, .$$

Its estimate $\hat{\mathbf{x}}_{k-3}^{(5)}$ for the time-step $k-3$ is not a member of the $\kappa = 2$ best estimates at this time-step. Discarding all branches that do not pass through nodes $n_i \in \bar{\mathcal{N}}_{k-3}$ would prevent hybrid estimation from considering an extension to the trajectory estimate $\hat{X}_{k-1}^{(2)}$. In order to avoid such an undesirable pruning of the hypothesis tree, we adopt a slightly different strategy for selecting the initial estimates for N-step hybrid estimation. Instead of treating κ as the fixed upper bound on the number of initial estimates, we interpret κ as the *upper lower bound* and restart N-step hybrid estimation from *at least* κ estimates at the time-step $k-N$. For this purpose, we analyze the leading set of trajectory estimates $\{\hat{X}_{k-1}^{(1)}, \ldots, \hat{X}_{k-1}^{(\eta)}\}$ at the time-step $k-1$ to obtain their hybrid state estimates for the time-step $k-N$. Let us use the following notation

$$\hat{X}_{k\backslash k-N}^{(\nu)} = \hat{\mathbf{x}}_{k-N}^{(\iota)} \tag{4.127}$$

to refer to the hybrid state estimate at a particular time-step $k-N$ of a trajectory estimate

$$\hat{X}_k^{(\nu)} = \{\ldots, \hat{\mathbf{x}}_{k-N}^{(\iota)}, \ldots, \hat{\mathbf{x}}_k^{(\nu)}\} \, . \tag{4.128}$$

Then, we can express the set of hybrid state estimates at the time-step $k-N$. This set is defined by the leading set of trajectory estimates $\{\hat{X}_{k-1}^{(1)}, \ldots, \hat{X}_{k-1}^{(\eta_{k-1})}\}$, as

$$\{\hat{X}^{(1)}_{k-1\backslash k-N} \cup \ldots \cup \hat{X}^{(\eta_{k-1})}_{k-1\backslash k-N}\} =: \bar{\mathcal{X}}_{k-N} .\qquad (4.129)$$

Some of the trajectory estimates trace back to common estimates at time-step $k-N$. This implies that the resulting set $\bar{\mathcal{X}}_{k-N}$ (4.129) can contain up to η_{k-1} estimates. We use this set to extract the rank ξ of the estimate $\hat{\mathbf{x}}^{(\xi)}_{k-N} \in \bar{\mathcal{X}}_{k-N}$ with the lowest likelihood $b^{(\xi)}_{k-N}$:

$$\xi = \underset{i \,:\, \hat{\mathbf{x}}^{(i)}_{k-N} \in \bar{\mathcal{X}}_{k-N}}{\arg\min} \; b^{(i)}_{k-N} .\qquad (4.130)$$

This rank indicates, whether we have to extend the set $\bar{\mathcal{N}}_{k-N}$ to include additional nodes ($\xi > \kappa$), or not ($\xi \le \kappa$). More generally, the function

truncate-search-tree-and-re-initiate($\cdot$) ,

which truncates the tree for N-step hybrid estimation (see Table 4.4), uses the following adaptive set $\bar{\mathcal{N}}_{k-N}$:

$$\bar{\mathcal{N}}_{k-N} = \mathbf{nodes}(\{\hat{\mathbf{x}}^{(1)}_{k-N},\ldots,\hat{\mathbf{x}}^{(\kappa_{\max})}_{k-N}\}), \; \text{with} \; \kappa_{\max} = \max(\xi,\kappa) .\quad (4.131)$$

Figure 4.24 visualizes this concept ($N = 3, \kappa = 2$). The $\eta = 3$ most likely trajectory estimates at time-step $k-1$ are $\{\hat{X}^{(1)}_{k-1}, \hat{X}^{(2)}_{k-1}, \hat{X}^{(3)}_{k-1}\}$. These trajectory estimates trace back to

$$\hat{X}^{(1)}_{k-1\backslash k-3} = \hat{\mathbf{x}}^{(2)}_{k-3}, \quad \hat{X}^{(2)}_{k-1\backslash k-3} = \hat{\mathbf{x}}^{(2)}_{k-3}, \quad \hat{X}^{(3)}_{k-1\backslash k-3} = \hat{\mathbf{x}}^{(5)}_{k-3}$$

so that we obtain the set

$$\bar{\mathcal{X}}_{k-3} = \{\hat{\mathbf{x}}^{(2)}_{k-3}, \hat{\mathbf{x}}^{(5)}_{k-3}\} .$$

The estimate $\hat{\mathbf{x}}^{(5)}_{k-3}$ represents the estimate with the lowest conditional probability of $\bar{\mathcal{X}}_{k-3}$ and specifies $\xi = 5$. As a consequence, we obtain $\kappa_{\max} = 5$ and utilize the extended set

$$\bar{\mathcal{N}}_{k-3} = \mathbf{nodes}(\{\hat{\mathbf{x}}^{(1)}_{k-3},\ldots,\hat{\mathbf{x}}^{(5)}_{k-3}\})$$

to restart N-step hybrid estimation for the time-step k.

Typically, real-world systems lead to hybrid models with a large number of modes. This leads to a heavily branching hypothesis tree. As a consequence, it is sometimes even computationally infeasible to apply N-step hybrid estimation with a moderately large window-size for real-time estimation so that we are forced to use N-step hybrid estimation with the smallest possible window-size of $N = 1$. The following section deals with this specific entity of N-step hybrid estimation in more detail.

Table 4.4. Pseudo-code for N-step hybrid estimation.

---

function initiate-N-step-HE-problem($\mathcal{CA}$, N, κ)
  returns initiated and main search problem
  HE-problem $\leftarrow$ initiate-HE-problem($\mathcal{CA}$)
  window-size[application-data[*HE-problem*]] $\leftarrow$ N
  kappa[application-data[*HE-problem*]] $\leftarrow$ κ
  return *HE-problem*

function N-step-hybrid-estimation($\mathbf{u}_k$, $\mathbf{y}_{c,k}$, $\eta_{k,\max}$, *HE-problem*)
  returns the leading set of sub-optimal estimates for time-step k
  $\{\mathcal{X}_k$, *HE-problem*$\}$ $\leftarrow$ hybrid-estimation($\mathbf{u}_k$, $\mathbf{y}_{c,k}$, $\eta_{k,\max}$, *HE-problem*)
  HE-problem $\leftarrow$ truncate-search-tree-and-re-initiate(*HE-problem*)
  return $\{\mathcal{X}_k$, *HE-problem*$\}$

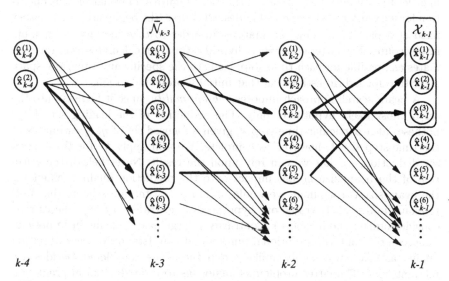

Fig. 4.24. Revised concept for the tree truncation of N-step hybrid estimation.

1-Step Hybrid Estimation - Beam Search

N-step hybrid estimation with the window-size $N = 1$ represents a special case where the high-level search problem degenerates to *beam search*. At a time-step k, we take the leading set (the *beam*) of fringe estimates

$$\{\hat{\mathbf{x}}_{k-1}^{(1)}, \ldots, \hat{\mathbf{x}}_{k-1}^{(\eta_{k-1})}\}$$

for the previous time-step $k - 1$ and deduce their leading successors

$$\{\hat{\mathbf{x}}_{k}^{(1)}, \ldots, \hat{\mathbf{x}}_{k}^{(\eta_k)}\}$$

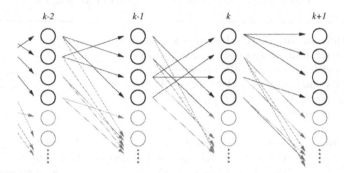

Fig. 4.25. Beam search (1-step hybrid estimation).

in order to estimate the current hybrid state. Figure 4.25 visualizes this operation for a fixed *beam-size* of $\eta = 4$ nodes. Of course, the beam-size can be either fixed at a particular value, or adaptive in order to take resource constraints into account. The latter case, where hybrid estimation calculates as many estimates as possible for each time-step, leads to the specific any-time/any-space algorithm for hybrid estimation that initially presented in [55].

A beam search approach imposes *sub-optimality*, since it focuses onto the set of most likely hypotheses only. The deduction of the leading set of hypotheses at a particular time-step k, given a focused set of η_{k-1} estimates at the time-step $k - 1$, however, is complete. We still apply A* for the successor deduction, so that we can rely upon the result. Nevertheless, the selection of the beam-size η_k has impact onto the prediction quality. Whenever the beam-size is too small, we would obtain an estimation scheme that fails to examine less likely trajectory hypotheses (e.g. trajectory hypotheses that consider unlikely fault modes). Contrarily, a large beam-size might impose an unnecessarily high computational burden and can deteriorate the estimation quality as too many, highly unlikely, hypotheses are considered, besides few relevant ones. The latter problem is analogous to the reduction in prediction quality of multiple-model estimation algorithms, whenever too many models are taken into consideration [68]. As a consequence, it is advisable to neither use too few estimates, nor to blindly deduce estimates until the computational resources are used up. Nevertheless, the 'optimal' beam-size depends upon the complexity of the model and the operational condition and is difficult to judge a priori.

The remaining part of this section is devoted to developing a suitable criteria for beam-size control. For this purpose, we first reconsider and interpret the estimation result. Let η_k denote the the number of estimates for trajectory hypotheses at the time-point k that will be used as the beam for the following estimation step. The leading set of trajectory hypotheses at time-step k has the associated trajectory estimates

$$\{\hat{X}_k^{(1)}, \hat{X}_k^{(2)}, \ldots, \hat{X}_k^{(\eta_k)}\}.$$

This set of trajectory estimates defines the set of hybrid state estimates

$$\mathcal{X}_k = \{\hat{\mathbf{x}}_k^{(1)}, \hat{\mathbf{x}}_k^{(2)}, \ldots, \hat{\mathbf{x}}_k^{(\eta_k)}\} \tag{4.132}$$

at their fringe.

One way to interpret this result is to take the hybrid state of the leading hypothesis $(\hat{\mathbf{x}}_k^{(1)})$ and utilize it as the the approximation of the hybrid estimate at the time-step k. This is in spirit of the *maximum a posterior (MAP)* approach of some multiple-model estimation schemes. However, such an approach can be misleading as we can see in the Table 4.5. The table lists a hypothesis 1 at the mode $\mathbf{m}_1$. Its likelihood of 0.4 suggests that the mode $\mathbf{m}_1$ is the most likely mode at the time-step k. The hypotheses 2 and 3 describe estimates with the mode $\mathbf{m}_2$. Together, they represent a larger portion of the probability space (46%), than hypothesis 1 (40%). Based on the 4 hypotheses only, one would conclude that the most likely mode at the time-step k is $\mathbf{m}_2$ (likelihood 0.46), followed by mode $\mathbf{m}_1$ (likelihood 0.40) and mode $\mathbf{m}_3$ (likelihood 0.14). A∗ search, as the underlying deduction mechanism for the estimates $\{\hat{\mathbf{x}}_k^{(1)}, \ldots, \hat{\mathbf{x}}_k^{(\eta_k)}\} = \mathcal{X}_k$, guarantees that the set of estimates $\mathcal{X}_k$ is complete, given the previous estimates $\{\hat{\mathbf{x}}_{k-1}^{(1)}, \ldots, \hat{\mathbf{x}}_{k-1}^{(\eta_k-1)}\} = \mathcal{X}_{k-1}$. Any continuation of the search adds less likely estimates, but cannot change the the ordering among the estimates of $\mathcal{X}_k$. As a consequence, we can be sure that $\hat{X}_k^{(1)}$, with the fringe estimate $\hat{\mathbf{x}}_k^{(1)}$, represents the leading *trajectory estimate*. However, the fringe estimates $\mathcal{X}_k = \{\hat{\mathbf{x}}_k^{(1)}, \ldots, \hat{\mathbf{x}}_k^{(\eta_k)}\}$ *as a whole* encode the hybrid state estimate at the time-step k. Thus, in order to obtain the overall hybrid state estimate, we have to combine the fringe estimates $\mathcal{X}_k$ according to (4.74)-(4.79) to obtain the most likely mode, the continuous estimate given the most likely mode, or the overall continuous estimate. Now, $\mathcal{X}_k$ represents the truncated set of fringe estimates. This raises the question whether the set is representative enough. For example, with respect to the mode prediction: is it possible to guarantee that additional estimates do not change the prediction of the most likely mode anymore? Recall the example above (Table 4.5). Mode $\mathbf{m}_2$ is selected as the most likely mode and the mode $\mathbf{m}_1$ is second. However, the difference in their likelihood is only 0.06 and it is possible that neglected hypotheses with mode $\mathbf{m}_1$ could reverse this ranking. This

Table 4.5. Posterior mode probabilities at time-step k.

hypothesis	mode $(\hat{\mathbf{x}}_{d,k}^{(i)})$	$\bar{b}_k^{(i)}$	probability $(b_k^{(i)})$
1	$\mathbf{m}_1$	0.20	0.40
2	$\mathbf{m}_2$	0.15	0.30
3	$\mathbf{m}_2$	0.08	0.16
4	$\mathbf{m}_3$	0.07	0.14

can be seen from the values of the unnormalized likelihoods $\bar{b}_k^{(i)}$. Our search mechanism guarantees that

$$\bar{b}_k^{(i)} \leq \bar{b}_k^{(\eta_k)}, \text{ whenever } i > \eta_k .$$

As a consequence, it is possible that there exists a fifth hypothesis $\hat{X}_k^{(5)}$ with a fringe state at mode $\mathbf{m}_1$ and an unnormalized likelihood $\bar{b}_k^{(5)} < 0.07$, for example, $\bar{b}_k^{(5)} = 0.05$. It is easy to see that this estimate has significant impact upon the mode predication and revises the estimate for the most likely mode, because $\bar{b}_k^{(5)} > (\bar{b}_k^{(2)} + \bar{b}_k^{(3)}) - \bar{b}_k^{(1)} = 0.03$. This indicates that we cannot guarantee a correct mode prediction given the first 4 hypotheses only! It is not just that single hypotheses can revert mode predictions, it is also possible that several hypotheses with marginal $\bar{b}_k$ add up to a level that changes the overall mode prediction.

A correct mode prediction is surely a desirable property for hybrid estimation. Therefore, we will use correctness of mode prediction, given the previous estimates $\mathcal{X}_{k-1}$, as the criteria that controls the beam-size of 1-step hybrid estimation. In the following we use the notation $\mathbf{m}^{(j)}$ to denote modes of the leading hybrid state estimates $\mathcal{X}_k = \{\hat{\mathbf{x}}_k^{(1)}, \ldots, \hat{\mathbf{x}}_k^{(\eta_k)}\}$, with

$$\hat{\mathbf{x}}_{d,k}^{(i)} \in \{\mathbf{m}^{(1)}, \mathbf{m}^{(2)}, \ldots, \mathbf{m}^{(\rho)}\}, \quad i = 1, \ldots, \eta_k .$$

The superscript j of $\mathbf{m}^{(j)}$ also denotes the rank of the mode, given the set of hybrid state estimates $\mathcal{X}_k$, which represents the leading successors of $\mathcal{X}_{k-1}$. This rank is based on the mode's associated unnormalized likelihood $\bar{b}_k(\mathbf{m}^{(j)})$ that aggregates the values of $\bar{b}_k^{(i)}$ of all hypotheses $\hat{\mathbf{x}}_k^{(i)}$ at mode $\mathbf{m}^{(j)}$, more precisely:

$$\bar{b}_k(\mathbf{m}^{(j)}) := \sum_{i \,:\, \hat{\mathbf{x}}_{d,k}^{(i)}=\mathbf{m}^{(j)}} \bar{b}_k^{(i)} .$$

In order to determine whether a mode $\mathbf{m}^{(j)}$ can become more likely then the leading mode $\mathbf{m}^{(1)}$, we need a bound for the maximal number of possible hypotheses with mode $\mathbf{m}^{(j)}$ at time-step k. We can obtain this information from the transition topology, in particular the transition threads, of the underlying cPHA model. The topology of the transition threads determines whether an estimate $\hat{\mathbf{x}}_{k-1}^{(i)} \in \mathcal{X}_{k-1}$ can lead to one, or to several hypotheses $\hat{\mathbf{x}}_k^{(\iota)}$ with the mode $\mathbf{m}^{(j)}$. For an example, recall Fig. 4.20. The transition threads for the transitions out of mode m_1 specify that an estimate $\hat{\mathbf{x}}_{k-1}^{(i)}$ with mode $\hat{\mathbf{x}}_{d,k-1}^{(i)} = m_1$ can lead to at most two hypotheses with mode m_1, to one hypothesis with mode m_2, and to one hypothesis with mode m_3. These upper bounds for hypotheses are key to decide, whether a mode $\mathbf{m}^{(j)}$ ($2 \geq j \geq \rho$) can become more likely than the leading mode $\mathbf{m}^{(1)}$. First, we determine the upper bound γ_j for hypotheses at the mode $\mathbf{m}^{(j)}$ by evaluating the transition treads out of the mode $\hat{\mathbf{x}}_{d,k-1}^{(i)}$ of the estimates $\hat{\mathbf{x}}_{k-1}^{(i)} \in \mathcal{X}_{k-1}$. Second, we determine the number $\nu_j \geq 1$ of estimates with mode $\mathbf{m}^{(j)}$ that were deduced so far. The difference $\gamma_j - \nu_j$ provides an upper bound for the number of unconsidered hypotheses with the mode $\mathbf{m}^{(j)}$. Every unconsidered hypothesis can have an associated unnormalized likelihood less, or at most equal to, $\bar{b}_k$ of the least likely hypothesis of $\mathcal{X}_k$. As a consequence, the unconsidered hypotheses at mode $\mathbf{m}^{(j)}$ can lead to an unnormalized likelihood for mode $\mathbf{m}^{(j)}$ of at most

$$\bar{b}_{\max}(\mathbf{m}^{(j)}) = \bar{b}_k(\mathbf{m}^{(j)}) + (\gamma_j - \nu_j)\bar{b}_k^{(\eta_k)} . \tag{4.133}$$

Whenever the condition

$$\bar{b}_k(\mathbf{m}^{(1)}) \geq \bar{b}_{\max}(\mathbf{m}^{(j)}) \quad \text{for all} \quad j = 2, \ldots, \rho \tag{4.134}$$

holds, we can be sure that the mode $\mathbf{m}^{(1)}$ represents the most likely mode.

We can simplify this criteria dramatically whenever we ensure that the underlying cPHA defines *at most one* transition $\mathbf{m}_i \to \mathbf{m}_j$ for all modes $\mathbf{m}_i, \mathbf{m}_j \in \mathcal{X}_d$. This modeling assumption implies that η_{k-1} estimates at time-step $k - 1$ can lead to at most $\gamma = \eta_{k-1}$ hypotheses with a particular mode $\mathbf{m}^{(j)}$ at time-step k (each hypothesis $\hat{\mathbf{x}}_{k-1}^{(i)}$ has at most one successor with mode $\mathbf{m}^{(j)}$). This upper bound is independent of the particular mode $\mathbf{m}^{(j)}$ under consideration. As a consequence, we only need to consider the leading two modes $\mathbf{m}^{(1)}$ and $\mathbf{m}^{(2)}$, and check

$$\bar{b}_k(\mathbf{m}^{(1)}) \geq \bar{b}_k(\mathbf{m}^{(2)}) + (\eta_{k-1} - 1)\bar{b}_k^{(\eta_k)} . \tag{4.135}$$

According to the cPHA model under investigation, we apply either (4.135) or (4.133)-(4.134) to adapt the fringe size η_k. However, we still terminate estimation for a particular time-step whenever we run out of the computational resources for hybrid estimation[9]. This maintains the desirable any-time/any-space property, even at the risk that we might sometimes fail to formally guarantee a correct mode prediction. The criteria (4.133)-(4.134) or (4.135) represent *sufficient* conditions that might be too strong anyhow. So that from a practical point of view, their infrequent violation has minor impact, compared to not being able to perform hybrid estimation in real-time. The pseudo-code in Table 4.6 summarizes this particular instance of our class of focused hybrid estimation algorithms and completes our presentation on these core algorithms for our hybrid estimation paradigm.

4.6 Unknown Mode and Filter Decomposition

The mode estimation scheme as defined above, as well as the standard multiple model estimation algorithms, assume that the system exhibits a mode of

---

[9] The pseudo-code also provides means to specify application-specific minimal and maximal numbers for the fringe size, $\eta_{k,\min}$ and $\eta_{k,\max}$, respectively.

Table 4.6. Pseudo-code for 1-step hybrid estimation (beam search).

---

function initiate-beam-search-HE-problem($\mathcal{CA}$)
  returns initiated and main search problem
  HE-problem ← initiate-HE-problem($\mathcal{CA}$)
  return *HE-problem*

function beam-search-hybrid-estimation($\mathbf{u}_k$, $\mathbf{y}_{c,k}$, $\eta_{k,\max}$, $\eta_{k,\min}$, *HE-problem*)
  returns the leading set of sub-optimal estimates for time-step k
  time-step[application-data[*HE-problem*]] ← +1
  input-value[application-data[*HE-problem*]] ← add-value($\mathbf{u}_k$)
  observation[application-data[*HE-problem*]] ← add-value($\mathbf{y}_{c,k}$)
  agenda[*HE-problem*] ← solutions[*HE-problem*]
  solutions[*HE-problem*] ← { }
  do progressive-best-first-search(*HE-problem*)
    until ((guaranteed-mode-prediction(*HE-problem*) **and**
        count(solutions[*HE-problem*]) $\geq \eta_{k,\min}$) **or**
        resources-used-up?() **or** count(solutions[*HE-problem*]) $= \eta_{k,\max}$)
  return {hybrid-estimate(solutions[*HE-problem*]), HE-problem}

---

operation within the set of modes that is captured by the model. Mathematical models, however, always represent approximations of the real world and many recent efforts in estimation and fault detection and isolation (FDI) were devoted to building *robust* estimation/diagnosis algorithms that can cope with unavoidable inaccuracy and incompleteness of the model [39, 26]. However, unanticipated failures do occur in real world systems and robust estimation and diagnosis methods might fail to detect such an unusual operational condition or can perform in an unexpected way. As a consequence, it is desirable to extend the estimation and diagnosis capability so that it can cope with and identify an *unknown mode* of operation.

The concept of the *unknown mode* is central to discrete model-based diagnosis [46]. Its underlying concept of constraint suspension [28] allows the diagnosis of systems where no assumption is made about the behavior of one or several components of the system. In this way, model-based diagnosis schemes, such as the *General Diagnostic Engine (GDE)* [29] or *Sherlock* [30], capture unspecified and unforeseen behaviors of the system by considering an *unknown mode* that does not impose any constraints on the system's variables.

We first demonstrated this principle for our hybrid estimation scheme in [54, 56], where we introduced the hybrid systems pendant of the unknown mode, together with a novel on-line filter deduction and decomposition capability. The following section represents the consequent refinement of these novel extensions to hybrid estimation that enable the estimator to continue its operation in the presence of unknown behavioral modes, as well as it improves filtering performance in terms of the filtering quality, on-line filter deduction, and the filter execution time.

4.6.1 Unknown Mode

The treatment of the unknown mode for hybrid systems is similar as for discrete model-based diagnosis. We extend the cPHA model with an additional mode that does not impose any constraint upon the continuous variables of the automaton. In other words, the automaton model provides an empty set of equations

$$F(\mathbf{m}_?) = \{\}$$

for the unknown mode $\mathbf{m}_?$. Our cPHA compiler appends an unknown mode to each PHA $\mathcal{A}_i$ of the cPHA and embeds the unknown mode in terms of unlikely transitions from each nominal mode to the unknown mode. Figure 4.26 illustrates this for a single PHA component with 3 nominal modes. Each self-loop for the nominal modes $m_1, \ldots, m_3$ contains a low probability thread to the unknown mode $m_?$. This models the low probability chance that the system enters an unanticipated mode of operation, regardless of the current operational mode. Not shown are the reverse transitions from the unknown mode back to the nominal modes. The unknown mode has one unguarded transition with equally likely threads to itself and all nominal modes. This transition ensures that hybrid estimation can recover from an unknown mode detection. It also provides a suitable reset functionality that restarts hybrid estimation from an equally distributed set of modes as soon as the truncated set of possible hypotheses does not explain the observations anymore.

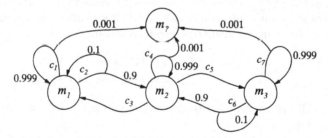

Fig. 4.26. Transition graph for a PHA with 3 nominal modes $m_1, \ldots, m_3$ and an unknown mode $m_?$.

In terms of hybrid estimation, we can utilize the transition probabilities to calculate the prior probability for a trajectory hypothesis with a fringe estimate at the unknown mode. However, we cannot perform any filtering operation since we do not provide a mathematical model for the unknown mode. We use a generic *no-prediction* instead and assume that any observation

$\mathbf{y}_{c,k}$ is conform with it[10]. We express this fact by taking the upper bound $\bar{P}_{\mathcal{O}} = 1.0$ for the observation function. This assumption enables us to calculate the posterior probability for an unknown mode hypothesis. Thus, even if we are unable to provide an update for the continuous estimate, we can still rate the associated trajectory hypothesis against other hypotheses with fringe estimates at nominal modes.

The unknown mode $\mathbf{m}_? = [\mathbf{m}_{1?}, \ldots, \mathbf{m}_{\zeta?}]^T$ specifies an unknown behavioral situation for all ζ components $\{\mathcal{A}_1, \ldots, \mathcal{A}_\zeta\}$ of the cPHA. This mode is surely very helpful to achieve a *fail-safe* operation of the hybrid estimation procedure. However, it does not identify those parts of the multi-component system that are responsible for this unspecified operation. Hybrid estimation is doing a good job in discriminating the modes of operation for the individual components. It is desirable to have the same capability for the unknown mode. This would enable us to *isolate* (diagnose) those components that operate at an unanticipated mode of operation, and maintains the hybrid estimation functionality for the remaining components. The following section presents a filter decomposition scheme that will serve as the basis for the unknown mode diagnosis capability and the (degraded) hybrid estimation functionality in the presence of unknown behavioral modes of individual components.

4.6.2 Filter Decomposition

Hybrid estimation deduces the trajectory hypotheses by repeatedly generating possible modes $\hat{\mathbf{x}}_{d,k}^{(j)}$ with consecutive filtering. The unknown mode modification of the cPHA can lead to mode hypotheses, where one or several components are at $\mathbf{m}_{\nu?}$. Such a mode hypothesis, however, leads to an incomplete set of equations $F(\hat{\mathbf{x}}_{d,k}^{(j)})$ since components with mode $\mathbf{m}_{\nu?}$ do not impose any constraint upon their continuous state and I/O variables. As a consequence, we cannot deduce the mathematical model of form

$$F(\hat{\mathbf{x}}_{d,k}^{(i)}) \quad \Longrightarrow \quad \begin{aligned} \mathbf{x}_{c,k+1} &= \mathbf{f}(\mathbf{x}_{c,k}, \mathbf{u}_{c,k}) + \mathbf{v}_{cx,k} \\ \mathbf{y}_{c,k} &= \mathbf{g}(\mathbf{x}_{c,k}, \mathbf{u}_{c,k}) + \mathbf{v}_{cy,k} \end{aligned} \tag{4.136}$$

for the overall system and fail to derive the extended Kalman filter for the estimation step.

Let us consider a 3-component example as shown in Fig. 4.27 with the cPHA

$$\begin{aligned} \mathcal{CA} = \langle &\mathcal{A}_1 \parallel \mathcal{A}_2 \parallel \mathcal{A}_3, \{u_{c1}, u_{d1}, u_{d2}\}, \{y_{c1}, y_{c2}\}, \\ &\{v_{c1}, v_{c2}, v_{c3}, v_{c4}, v_{c5}\}, N \rangle \,, \end{aligned} \tag{4.137}$$

which is composed of the following component automata

---

[10] The unknown mode captures all possible unmodeled situations, therefore, we have to assume that its no-prediction is more compatible with the observation, than any of the more precise predictions of the nominal modes.

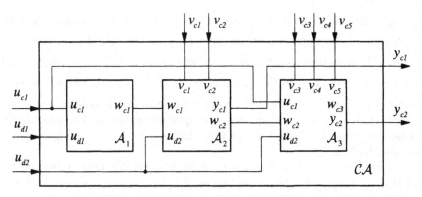

Fig. 4.27. Example cPHA composed of three PHAs.

$$\mathcal{A}_1 = \langle \{x_{d1}\}, \{u_{d1}, u_{c1}, w_{c1}\}, F_1, T_1, X_{01}, \{m_{11}, m_{12}\}...\rangle$$
$$\mathcal{A}_2 = \langle \{x_{d2}, x_{c1}\}, \{u_{d2}, w_{c1}, w_{c2}, y_{c1}, v_{c1}, v_{c2}\}, F_2, T_2,$$
$$X_{02}, \{m_{21}, m_{22}, m_{23}\}...\rangle \qquad (4.138)$$
$$\mathcal{A}_3 = \langle \{x_{d3}, x_{c2}, x_{c3}\}, \{u_{d2}, u_{c1}, w_{c2}, w_{c3}, y_{c2}, v_{c3}, v_{c4}, v_{c5}\},$$
$$F_3, T_3, X_{03}, \{m_{31}, m_{32}, m_{33}\}...\rangle \ .$$

As an example, we assume that the set-valued functions F_1, F_2 and F_3 provide the following sets of equations for the mode $\hat{\mathbf{x}}_{d,k}^{(i)} = [m_{11}, m_{21}, m_{31}]^T$

$$F_1(m_{11}) = \{u_{c1} = 2.0 \ w_{c1}\}$$
$$F_2(m_{21}) = \{x_{c1,k+1} = 0.95 \ x_{c1,k} + w_{c1,k} + v_{c1,k},$$
$$w_{c2} = 2.0 \ x_{c1},$$
$$y_{c1} = w_{c2} + v_{c2}\}$$
$$F_3(m_{31}) = \{x_{c2,k+1} = x_{c3,k} + 0.2 \ w_{c2,k} + v_{c3,k}, \qquad (4.139)$$
$$x_{c3,k+1} = -0.63 \ x_{c2,k} + 1.6 \ x_{c3,k} + 0.1 \ u_{c1,k} + v_{c4,k},$$
$$w_{c3} = 0.5 \ x_{c2} + 0.1 \ x_{c3}$$
$$y_{c2} = w_{c3} + v_{c5}\} \ .$$

Symbolic manipulation of this raw model for the mode $[m_{11}, m_{21}, m_{31}]^T$ derives the discrete-time model:

$$x_{c1,k+1} = 0.95 \ x_{c1,k} + 0.5 \ u_{c1,k} + v_{c1,k}$$
$$x_{c2,k+1} = 0.4 \ x_{c1,k} + x_{c3,k} + v_{c3,k}$$
$$x_{c3,k+1} = -0.63 \ x_{c2,k} + 1.6 \ x_{c3,k} + 0.1 \ u_{c1,k} + v_{c4,k} \qquad (4.140)$$
$$y_{c1,k} = 2.0 \ x_{c1,k} + v_{c2,k}$$
$$y_{c2,k} = 0.5 \ x_{c2,k} + 0.1 \ x_{c3,k} + v_{c5,k} \ .$$

This model, together with the quantification of the noise inputs $\mathbf{v}_c$, is the basis for the extended Kalman filter that we use to provide the continuous state estimate for the hybrid estimate $\hat{\mathbf{x}}_k^{(i)}$, as well as the associated value for the modified hybrid observation function $\bar{P}_{\mathcal{O},k}^{(i)}$ for the hybrid estimator.

Let us now assume that the system is at the mode $\hat{\mathbf{x}}_{d,k}^{(j)} = [m_?, m_{21}, m_{31}]^T$ where component $\mathcal{A}_1$ is in the *unknown mode* $(\hat{\mathbf{x}}_{d1,k}^{(j)} = m_?)$. The set of equations $F(\hat{\mathbf{x}}_{d,k}^{(j)})$ fails to include any constraint that links the input u_{c1} with the I/O variable w_{c1}. As a consequence, our symbolic solver fails to provide the mathematical model of the form (4.136) that is the basis for the appropriate extended Kalman filter.

Nevertheless, an analysis of the PHA interconnection (Fig. 4.28) and the fact that y_{c1} represents the measurement of w_{c2} reveals that we can still estimate component $\mathcal{A}_3$ by its output y_{c2} and the observation y_{c1} that measures its input w_{c2}.

This intuitive approach to estimate $\mathcal{A}_3$ utilizes a decomposition of the cPHA into two subsystems as shown in Fig. 4.29. The decomposition allows us

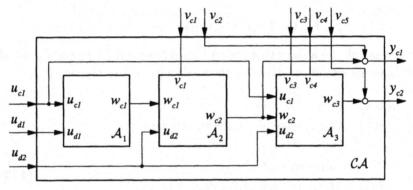

Fig. 4.28. Example cPHA extended with explicit sensor noise influence and causality (directionality) for the automata interconnections.

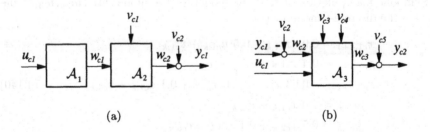

(a)   (b)

Fig. 4.29. cPHA decomposition.

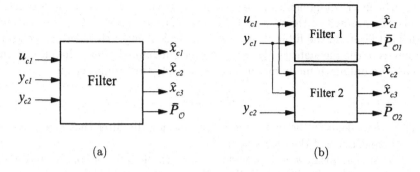

Fig. 4.30. Overall versus decomposed filter.

to treat the parts of the system independently by 2 concurrent filters as shown in Fig. 4.30b. The incomplete set of equations prevents us from calculating the first filter for the components $\mathcal{A}_1$ and $\mathcal{A}_2$, but it provides enough information to derive the second filter, which estimates the variables of $\mathcal{A}_3$. However, filter deduction has to account for the fact that we use the *measurement* y_{c1} of the input to $\mathcal{A}_3$ in replacement for its true value. This can be interpreted as having virtual additive noise at the component's input, as indicated in Fig. 4.29. For (extended) Kalman filters, this involves the following modification of the covariance matrix $\mathbf{Q}_3$ for the disturbances v_{c3}, and v_{c4} that act upon the continuous state variables of the automaton $\mathcal{A}_3$:

$$\tilde{\mathbf{Q}}_{f2} = \mathbf{b}_3 r_2 \mathbf{b}_3^T + \mathbf{Q}_3 , \qquad (4.141)$$

where r_2 denotes the variance of disturbance v_{c2} and $\mathbf{b}_3 = [0, \ 1]^T$ denotes the input vector of $\mathcal{A}_3$ with respect to w_{c2}. In general, when we replace a variable $w_c \in \mathbf{w}_c$ with its observation $y_c \in \mathbf{y}_c$ (assuming additive sensor noise), we obtain the associated input vector $\mathbf{b}_j$ by linearization. More specifically:

$$\mathbf{b}_{j,(k)} = \left. \frac{\partial \mathbf{f}_j}{\partial w_c} \right|_{\hat{\mathbf{x}}_{j,k-1}, \mathbf{u}_{c\,j,k-1}} , \qquad (4.142)$$

where $\mathbf{f}_j$ denotes the right-hand side of the difference equation for component $\mathcal{A}_j$, as well as $\hat{\mathbf{x}}_{j,k-1}$ and $\mathbf{u}_{c\,j,k-1}$ represent the hybrid state estimate and the continuous input for component $\mathcal{A}_j$ at the previous time-step, respectively.

The decomposition leads to a factorization of the probabilistic observation function

$$\bar{P}_{\mathcal{O}} = \prod_j \bar{P}_{\mathcal{O}j} , \qquad (4.143)$$

where $\bar{P}_{\mathcal{O}j}$ denotes the probabilistic observation function of the j'th concurrent filter. This factorization of $\bar{P}_{\mathcal{O}}$ allows us to calculate an upper bound for

$\bar{P}_{\mathcal{O}}$ whenever components of the system are in the unknown mode. We simply take the product of the values of the observation functions $\bar{P}_{\mathcal{O}\nu}$ over the well defined filters in the filter cluster. This is equivalent to considering the upper bounds of the inequalities $\bar{P}_{\mathcal{O}j} \leq 1$ for every undefined filter j. In our example with the unknown mode for the component $\mathcal{A}_1$ this would mean:

$$\bar{P}_{\mathcal{O}} \leq \bar{P}_{\mathcal{O}2} ,$$

where $\bar{P}_{\mathcal{O}2}$ denotes the observation function for the filter that estimates the continuous state of the component $\mathcal{A}_3$.

What remains now is to provide an algorithmic solution for the informally introduced decomposition and filter cluster deduction.

4.6.3 Graph-Based Decomposition and Filter Cluster Calculation

The decomposition of the system for a cPHA at the mode $\hat{\mathbf{x}}_d$ is has to be done according to the set of equations or *raw model* for the cPHA at the mode $\hat{\mathbf{x}}_{d,k}$

$$F(\hat{\mathbf{x}}_{d,k}) = F_1(\hat{\mathbf{x}}_{d1,k}) \cup \ldots \cup F_\zeta(\hat{\mathbf{x}}_{d\zeta,k}) . \tag{4.144}$$

This raw model specifies the relations among the variables and can be expressed as a graph so that graph-based structural analysis methods and decomposition can be applied. The methods described in the following build upon methods for causal analysis [83, 96], structural observability analysis [41, 87, 62] and graph decomposition [3].

The cPHA model does not impose a fixed causal structure that specifies the directionality of the automaton interconnections. The specification of the (exogenous) input variables $\mathbf{u}_c \in \mathbf{w}_c$ of the cPHA and the set of equations defines the causality implicitly. This increases the expressiveness of the modeling framework. Nevertheless, knowledge about the causal dependencies among the continuous variables of the model is very valuable for system analysis and filter synthesis tasks. We apply the bipartite-matching based algorithm of [83] for this purpose and obtain a *causal graph* that records the causal dependencies for a concurrent probabilistic automaton $\mathcal{CA}$ at a particular mode $\mathbf{x}_d$. We denote the causal graph by $\mathcal{CG}(\mathcal{CA}, \mathbf{x}_d)$ and sometimes omit the arguments when no confusion seems likely. Figure 4.31 shows the graph for the the illustrative 3 PHA example at the mode $\mathbf{x}_{d,k} = [m_{11}, m_{21}, m_{31}]^T$. Each vertex of the causal graph represents one equation $e_i \in F(\hat{\mathbf{x}}_{d,k})$ or an exogenous variable specification (e.g. u_{c1}) and is labeled by its *dependent variable*, which also specifies the outgoing edge. In the following, we will use the variable name to refer to the corresponding vertex in the causal graph. Vertices without incoming edges specify the *exogenous* (or independent I/O) variables.

The aim of our structural analysis is to obtain a set of non-overlapping subsystems that utilize known inputs and observed variables as inputs. For this purpose, we identify those variables that are directly observed through the measurements. We do this by traversing the causal graph from vertices

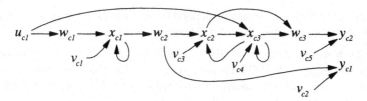

Fig. 4.31. Causal graph for the cPHA example.

of the observed variables $y_{ci} \in \mathbf{y}_c$ backwards to identify those vertices with dependent variables that can be fully specified in terms of the observation and the input variables $\mathbf{u}_c \cup \mathbf{v}_c$ of the cPHA. Those vertices with additional out-going edges (other than the edge that leads to the measurement), are selected as points, where we can slice the graph and remap it to a *virtual input*. This operation, when applied to our example (Fig. 4.31), infers that the variables w_{c2} and w_{c3} are directly observable via y_{c1} and y_{c2}, respectively. Only the vertex with dependent variable w_{c2} has an additional outgoing edge, thus we slice the graph at the edge $w_{c2} \to x_{c2}$ and remap this edge to a vertex w'_{c2} that denotes a copy of the measurement equation (in reversed causal direction). Finally we insert the edge to the *virtual input* y'_{c1} and the noise input v_{c2}. Figure 4.32 demonstrates this remapping operation.

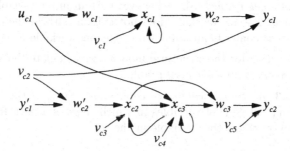

Fig. 4.32. Remapped causal graph for the cPHA example.

One can only estimate the *observable part* of the model. This makes it necessary to perform an observability analysis prior filter cluster synthesis so that non-observable parts of the model are excluded. We perform this analysis on a structural basis[11] and evaluate for every variable $z \in \mathbf{x}_c \cup \mathbf{w}_c$ whether

---

[11] We assume that a loss of observability is caused by a structural defect of the model (e.g. a stuck-at fault of a sensor that disconnects, in the causal graph, the measurement from the rest of the model). Otherwise, it would be necessary to perform an additional numerical observability test [91] as structural observability only provides a *necessary* condition for observability.

there exists at least one causally dependent output variable $y_c \in \mathbf{y}_c$ that can be used to estimate the value of z. More specifically:

Definition 4.1. We call a variable $z \in \mathbf{x}_c \cup \mathbf{w}_c$ of a cPHA $\mathcal{CA}$ at mode $\mathbf{x}_{d,k}$ *structurally observable* (SO) whenever it is directly observed, that is, $z \in \mathbf{y}_c$, or there exists at least one path in the causal graph $CG(\mathcal{CA}, \mathbf{x}_{d,k})$ that connects the variable z to an output variable $y_c \in \mathbf{y}_c$ of the $\mathcal{CA}$.

Estimation of the continuous state variables $\mathbf{x}_c$ (and, as a consequence, the other variables $\mathbf{w}_c$) is based on measurements $\mathbf{y}_c$ and the actuated inputs $\mathbf{u}_c$ that influence the state variables $\mathbf{x}_c$. Thus, structural observability alone is not yet enough, one also needs knowledge about the exogenous (independent) variables that act upon the state variables. This implies that no variable with unknown value should influence the state variables. Let us demonstrate this fact with the 3 PHA example at mode $\mathbf{x}_d = [m_?, m_{21}, m_{31}]^T$. The corresponding raw model omits any equation that relates the variable u_{c1} with w_{c1} because of the unknown mode of component 1. This leads to a causal graph $\tilde{C}\mathcal{G}$ (Fig. 4.33) that labels w_{c1} as exogenous. Thus, w_{c1} acts as an *unknown exogenous input* that influences the state variable x_{c1} and, as a consequence, prevents us from estimating it. Again, we identify variables that causally depend upon unknown exogenous variables in terms of a structural analysis of the causal graph, more specifically:

Definition 4.2. We call a variable $z \in \mathbf{x}_c \cup \mathbf{w}_c$ of a cPHA $\mathcal{CA}$ at mode $\mathbf{x}_{d,k}$ *structurally determined* (SD) whenever it is an input variable of the automaton, that is, $z \in \mathbf{u}_c$, or there does not exist a path in the causal graph $CG(\mathcal{CA}, \mathbf{x}_{d,k})$ that connects an unknown exogenous variable $u_e \notin \mathbf{u}_c$ with z.

As a preparation for the structural analysis we calculate the *strongly connected components* [3] of the causal graph:

Definition 4.3. A *strongly connected component (SCC)* of the causal graph CG is a maximal set SCC of variables in which there is a path from any one variable in the set to another variable in the set.

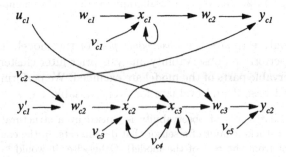

Fig. 4.33. Remapped causal graph for the cPHA example with unknown component $\mathcal{A}_1$.

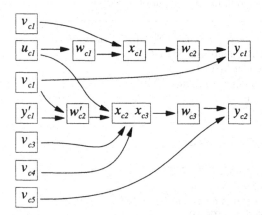

Fig. 4.34. Causal SCC graph for cPHA example.

Figure 4.34 shows the remapped causal graph for the 3 PHA example after grouping variables into strongly connected components. The tight interconnection among variables in a strongly connected component ensures that structural observability of variables in a strongly connected component follows directly from structural observability of at least one variable in the strongly connected component. Furthermore, a variable in a strongly connected component is structurally determined, if and only if all variables in the strongly connected component are structurally determined. These facts allow us to apply our structural analysis directly to strongly connected components and operate on the acyclic strongly connected component graph, that is, a causal graph without loops.

Our structural analysis algorithm determines structural observability and determination of a variable by traversing the strongly connected component graph backwards from the observed variables toward the inputs. Table 4.7 outlines the analysis of a strongly connected component with respect to structural observability and structural determination (SOD) in pseudo-code. In the course of this analysis we label non-exogenous strongly connected components with a tag that records their influence upon observed variables. This indexing scheme allows us to cluster the variables into non-overlapping subsets (Table 4.8). The direct relation between a variable, its determining equation, and the cPHA component that specified this equation leads to the component clusters sought for.

Each component cluster defines the observable and determined raw model for a subsystem of the cPHA. This raw model can be solved symbolically and provides the nonlinear system of difference equations, which is similar to (4.136) but has additional virtual inputs. This nonlinear system is the basis for the corresponding filter in the filter cluster. In this way we exclude the unobservable and/or undetermined parts of the overall system from estimation.

Table 4.7. Pseudo-code for structural observability and determination evaluation.

---

function determine-SOD-of-SCC(SCC, $\mathbf{u}_c$, $\mathbf{v}_c$, k)
  when SOD-undetermined?(SCC)
    if exogenous?(SCC)
      z_i ← independent-var(SCC)
      if $z_i \in \mathbf{u}_c \cup \mathbf{v}_c$
        SD(SCC) ← **True**
      else
        SD(SCC) ← **False**
    else
      $\mathcal{V}$ ← uplink-SCCs(SCC)
      loop for SCC_i in $\mathcal{V}$
        do determine-SOD-of-SCC(SCC_i, $\mathbf{u}_c$, $\mathbf{v}_c$, k)
      SO(SCC) ← **True**
      SD(SCC) ← all-uplink-SCCs-are-SD?($\mathcal{V}$)
      cluster-index(SCC) ← k ∪ cluster-indices($\mathcal{V}$)
    SOD-determined(SCC) ← **True**
  return Nil

Table 4.8. Pseudo-code for component clustering.

---

function component-clustering($\mathcal{CA}$, $\mathbf{x}_d$)
  returns a set of cPHA component clusters
  $\mathbf{y}_c$ ← observed-vars($\mathcal{CA}$)
  $\tilde{\mathcal{CG}}$ ← remap-causal-graph($\mathcal{CG}(\mathcal{CA}, \mathbf{x}_d)$, $\mathbf{y}_c$)
  $\mathbf{u}_c$ ← virtual-inputs($\tilde{\mathcal{CG}}$) ∪ input-vars($\mathcal{CA}$)
  $\mathbf{v}_c$ ← disturbance-inputs($\mathcal{CA}$)
  $\mathcal{CG}_{SCC}$ ← strongly-connected-component-graph($\tilde{\mathcal{CG}}$)
  $k \leftarrow 0$
  loop for SCC_i in output-SCCs($\mathcal{CG}_{SCC}$, $\mathbf{y}_c$)
    do determine-SOD-of-SCC(SCC_i, $\mathbf{u}_c$, k)
      $k \leftarrow k + 1$
  graph-clusters ← get-SOD-SSC-clusters($\mathcal{CG}_{SCC}$)
  return automaton-clusters($\mathcal{CA}$, *graph-clusters*)

4.6.4 Filter Cluster Deduction for Hybrid Estimation

The clustered filter that consists of several concurrent extended Kalman filters can be used instead of the overall filter as both provide the same expected value for the continuous state ($E\{\mathbf{x}_{c,k}\}$) for nominal modes of the cPHA. The filter cluster, however, is superior in terms of filter execution. This is due to the fact that the computational requirements for a Kalman Filter with n_x

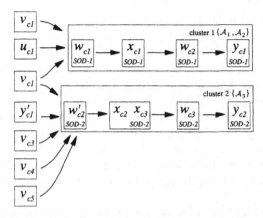

Fig. 4.35. Labeled and partitioned causal SCC graph for the 3 cPHA example.

state variables are approximately proportional to n_x^3 [45] so that the execution of several "small" filters outperforms the execution of a single "large" filter for the overall system. As a consequence, we do not limit decomposition to hypotheses with unknown modes alone, but use filter clusters also for nominal modes $\mathbf{m}_j \in \mathcal{X}_d$ of the cPHA.

The number of modes of hybrid models for a complex physical artifact can be very large. This prevents us from designing all filters a priori. However, deducing the same filter clusters over and over again would impose an unnecessarily large computational burden. Therefore, we *cache* a limited number of filters for re-use. This strategy represents a good compromise between the run-time cost of on-line deduction and the memory requirement of designing all filters a priori. Decomposition helps here as well. It significantly reduces the number of filter deductions since we can re-use the cached filters as *building blocks* for filter clusters. For example, the 3-PHA system (4.138) has $2 \times 3 \times 3 = 18$ possible modes. The deduction of the 18 possible filter clusters (Fig. 4.30b), however, requires only 9 filter deductions, $2 \times 3 = 6$ variations for filter 1 (components $\mathcal{A}_1$ and $\mathcal{A}_2$), and 3 variations for filter 2 (component $\mathcal{A}_3$).

We integrate the filter cluster deduction scheme in our hybrid estimation framework as follows. Continuous filtering represents the final part of the best-fist successor generation search (BFSG) problem. The first ζ steps of this underlying search problem deduce a suitable mode candidate $\hat{\mathbf{x}}_{d,k}^{(j)}$, according to the transition probabilities as shown in Fig. 4.16. A consecutive expansion of the associated node executes the continuous estimation. This step takes the hybrid estimate immediately after the transition $\hat{\mathbf{x}}_{k-1}^{'(j)} = \langle \hat{\mathbf{x}}_{d,k}^{(j)}, p_{c,k-1}^{'(j)} \rangle$, the inputs $\mathbf{u}_{c,k-1}, \mathbf{u}_{c,k}$, and the observation $\mathbf{y}_{c,k}$ and deduces a new estimate $\hat{\mathbf{x}}_k^{(j)}$. This involves the following operations: (a) retrieval of the raw model $F(\hat{\mathbf{x}}_{d,k}^{(j)})$, (b) the decomposition thereof, (c) filter cache retrieval or the application of an algebraic solver to deduce the subsets of difference and algebraic equations for each component cluster with consecutive filter deduction, and finally, (d) execution of the filter cluster that leads to the new estimate $\hat{\mathbf{x}}_k^{(j)}$. With respect to filter execution, we handle the unknown mode as follows. Whenever a state variable x_{cj} is unobservable and/or undetermined, we hold its mean at the last known estimate $\hat{x}_{cj}$ and increase its variance $\sigma_j^2 = p_{jj}$ by a constant factor. This reflects a decreasing confidence in the estimate $\hat{x}_{cj}$ and allows us to restart estimation whenever the variable becomes observable and determined again[12]. This operation ensures a defined behavior whenever some components of the system exhibit unanticipated behavior. Loss of the continuous estimation capability is limited to the impaired components and hybrid estimation continues for the fully specified components of the system.

Finally, we want to motivate our decision to perform the decomposition of the hybrid estimation problem on-line and at the continuous estimation level. For example, one could also decide to assume a fixed decomposition that divides the hybrid estimation problem into several sub-problems of smaller complexity. Nevertheless, taking the system-wide view makes sense as we will see below.

Let us extend the 3-PHA example with explicit sensor components. Component $\mathcal{A}_2$ only provides the internal I/O variable w_{c2} and $\mathcal{A}_3$ only provides w_{c3}. The observation of these variables, in terms of the output variables y_{c1} and y_{c2}, is subject to the dedicated sensor components $\mathcal{A}_4$ and $\mathcal{A}_5$, as shown in Fig. 4.36. The sensor components $\mathcal{A}_4$ and $\mathcal{A}_5$ model the observation of the internal variables, together with the impact of the sensor noise v_{c2} and v_{c5}. A simple model describes them in terms of two modes, operational and unknown. The operational modes m_{41} and m_{51} for the components $\mathcal{A}_4$ and $\mathcal{A}_5$, respectively, specify

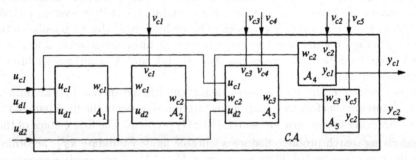

Fig. 4.36. Example cPHA composed of five PHAs, whereas two of them are dedicated sensor components.

---

[12] Whenever a state variable x_{cj} is directly observed we also can utilize an alternative approach suggested in [80] that restarts the estimator with the observed value, thus improving the observer convergence time.

$$\mathcal{A}_4 : \quad F_4(m_{41}) = \{y_{c1} = w_{c2} + v_{c2}\}$$
$$\mathcal{A}_5 : \quad F_5(m_{51}) = \{y_{c2} = w_{c3} + v_{c5}\} , \quad (4.145)$$

whereas the unlikely modes $m_{4?}$ and $m_{5?}$ are integrated as usual and capture all possible sensor faults.

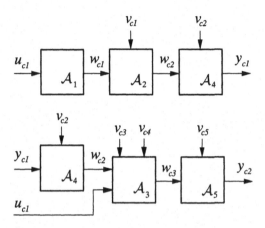

Fig. 4.37. Decomposed cPHA.

The nominal mode hypothesis $\hat{\mathbf{x}}'_{d,k-1} = [m_{11}, m_{21}, m_{31}, m_{41}, m_{51}]^T$ leads to a decomposition into two subsystems, similarly as above (Fig. 4.37). Contrarily, let us consider the mode $\hat{\mathbf{x}}'_{d,k-1} = [m_{11}, m_{21}, m_{31}, m_{4?}, m_{51}]^T$ that describes a sensor fault in component $\mathcal{A}_4$. This mode hypothesis implies that we cannot utilize the observation y_{c1} for estimation. As a consequence, we fail to decompose the system into two subsystems and hybrid estimation has to infer the state of the remaining components based on the observation y_{c2}. This mode dependency of the decomposition justifies our decision to assume that the model has no fixed decomposition topology. A fixed decomposition, for example, an estimator for the components $\{\mathcal{A}_1, \mathcal{A}_2, \mathcal{A}_4\}$ and an estimator for the components $\{\mathcal{A}_3, \mathcal{A}_4, \mathcal{A}_5\}$ could never properly consider the case, where $\mathcal{A}_4$ is at the unknown mode. Thus we cannot expect that a decomposed estimator captures the whole spectrum of possible behaviors that a system can exhibit. Our hybrid estimation scheme, however, uses a system-wide view to decide upon which mode hypotheses are of interest, and decomposes the continuous estimation task whenever a mode hypothesis admits a factorization.

5

Case Studies

This section demonstrates our hybrid estimation scheme on the basis of the illustrative 3-component cPHA and the BIO-Plex process automation example that were introduced above. We will put the emphasis on (a) comparing the variants of our focused hybrid estimation scheme (N-step hybrid estimation and 1-step hybrid estimation) and compare them to the interacting multiple model (IMM) algorithm, a prominent member of the class of multiple-model algorithms. The BIO-Plex process automation example allows us to demonstrate our proposed hybrid estimation scheme with a moderately complex multi-component system (8 components, 451,584 modes) and shows how the algorithm discriminates between various operational modes, fault modes, and unknown modes of operation of individual components. This section summarizes and extends the experiments on our hybrid estimation framework, that were previously given in [54, 55, 56].

5.1 Three Component Example

The three-component cPHA that we introduced above (4.137)-(4.139) is used first to compare the variants of our hybrid estimation scheme to multiple model estimation algorithms, in particular, an IMM- and a variable-structure IMM based hybrid estimator. As a reference, we use an "optimal" estimator that utilizes the correct mode information and performs the continuous estimation only.

The mathematical model for the modes m_{11}, m_{21} and m_{31} of the three components were given in (4.139). For the other modes of the cPHA components, we specify the variations in the equation-sets as follows:

$$F_1(m_{12}) = \{u_{c1} = -2.0 \; w_{c1}\}$$
$$F_2(m_{22}) = \{x_{c1,k+1} = 0.60 \; x_{c1,k} + w_{c1,k} + v_{c1,k}, \ldots\}$$
$$F_2(m_{23}) = \{x_{c1,k+1} = 1.01 \; x_{c1,k} + w_{c1,k} + v_{c1,k}, \ldots\}$$
$$F_3(m_{32}) = \{\ldots, x_{c3,k+1} = -0.80 \; x_{c2,k} + 1.6 \; x_{c3,k} + 0.1 \; u_{c1,k} + v_{c4,k}, \ldots\}$$
$$F_3(m_{33}) = \{\ldots, x_{c3,k+1} = -0.30 \; x_{c2,k} + 1.1 \; x_{c3,k} + 0.1 \; u_{c1,k} + v_{c4,k}, \ldots\} \,.$$
$$(5.1)$$

The discrete evolution of the mode is captured in terms of the transition graphs in Fig. 5.1. We use a transition scheme that is independent of the continuous state variables and the command inputs. The transition probabilities label the arcs in the digraphs. The disturbances $v_{c1}, \ldots, v_{c5}$ are assumed to be white, zero-mean Gaussian random sequences. Separating the disturbances into the state disturbances $\mathbf{v}_{cx} := [v_{c1}, v_{c3}, v_{c4}]^T$ and the measurement noise $\mathbf{v}_{cy} := [v_{c2}, v_{c5}]^T$, allows us to specify them by the covariance matrices

$$E\{\mathbf{v}_{cx,k}\mathbf{v}_{cx,k}{}^T\} = \begin{bmatrix} 0.4 & 0 & 0 \\ 0 & 0.5 & 0 \\ 0 & 0 & 0.3 \end{bmatrix}$$
$$(5.2)$$

$$E\{\mathbf{v}_{cy,k}\mathbf{v}_{cy,k}{}^T\} = \begin{bmatrix} 0.1 & 0 \\ 0 & 0.3 \end{bmatrix},$$

where in terms of Kalman filter nomenclature (4.3) $E\{\mathbf{v}_{cx,k}\mathbf{v}_{cx,k}{}^T\}$ denotes the covariance matrix $\mathbf{Q}$ for state disturbances and $E\{\mathbf{v}_{cy,k}\mathbf{v}_{cy,k}{}^T\}$ denotes the covariance matrix $\mathbf{R}$ of the measurement noise.

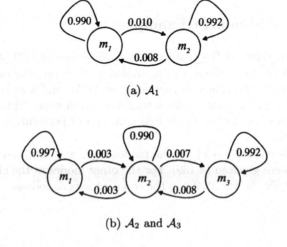

(a) $\mathcal{A}_1$

(b) $\mathcal{A}_2$ and $\mathcal{A}_3$

Fig. 5.1. Transition graphs for the three PHA example.

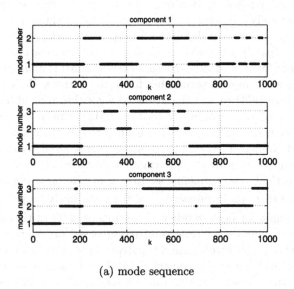

(a) mode sequence

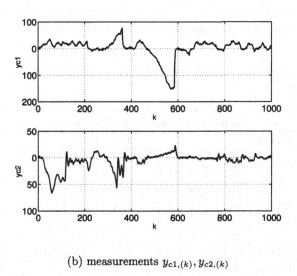

(b) measurements $y_{c1,(k)}, y_{c2,(k)}$

Fig. 5.2. Time-plots of the 3-PHA experiment.

The following algorithm analysis is based on a randomly generated simulation with $k_{max} = 5{,}000$ time-samples (Fig. 5.2 shows the mode sequence and measurements for the first 1,000 samples). We report the average relative estimation error of the algorithms to compare continuous filtering quality of the various algorithms. Mode estimation quality is reported in terms of the number of wrongly classified modes (single, double and triple mode estimation errors, relative to the overall number of estimations (k_{max}) in).

Table 5.1 (partly given also as Table I in [56]) summarizes the estimation results for the IMM, variable-structure IMM (vs-IMM) and our 1-step hybrid estimation algorithm with various but fixed fringe sizes η. We analyze both the basic algorithm (hMEη) and the extended algorithm that utilizes component clustering, as described above (c-hMEη). hME tests $m > \eta$ hypotheses in order to obtain the leading set of η estimates (i.e. it deduces and executes m extended Kalman filters). The columns average and maximal number of tested hypotheses provide this information. In terms of the estimator output for these particular experiments we take a maximum a posterior approach (MAP) and utilize the estimate $\hat{\mathbf{x}}_k^{(1)}$ of the most likely trajectory hypothesis $M_{d,k}^{(1)}$ as estimate for both, mode and continuous state at time-step k.

We also included a runtime comparison based on relative run-times for each algorithm under investigation (relative to the standard IMM algorithm). This is legitimate as we implemented IMM and variable-structure IMM as a subset of the overall cPHA estimation engine that is written in Common LISP. All estimation algorithms (IMM, vs-IMM, hME and c-hME) utilize the same concurrent filter bank data structures and calling mechanisms, thus we can also meaningfully compare the runtime performances.

Table 5.1. Algorithm comparison (a shortened version of this table was given as Table I in [56]).

Algorithm	rel. error e	mode errors [%]			correct mode [%]	fringe size	tested hypotheses		rel. runtime
		single	double	triple			average	max	
optimal	0.1100	-	-	-	100	1	1	1	-
IMM	0.1130	12.6	1.5	0.3	85.6	18	18	18	1.00
vs-IMM	0.1130	39.4	24.7	0.8	35.1	-	10.4	18	0.47
hME1	0.1534	41.7	19.8	5.9	32.6	1	2.1	18	0.06
hME2	0.1367	37.2	14.7	1.7	46.4	2	3.9	36	0.10
hME5	0.1173	20.5	7.0	0.4	72.1	5	10.3	70	0.24
hME10	0.1170	17.5	5.6	0.6	76.3	10	20.8	140	0.47
hME15	0.1174	17.9	4.7	0.5	76.9	15	31.9	222	0.76
hME20	0.1172	18.3	5.6	0.6	75.5	20	42.0	280	0.98
c-hME1	0.1455	42.9	19.6	6.4	31.1	1	1.9	18	0.05
c-hME2	0.1190	24.0	9.6	0.7	65.7	2	4.0	36	0.08
c-hME5	0.1169	17.1	4.4	0.2	78.3	5	10.4	70	0.19
c-hME10	0.1167	16.7	4.5	0.2	78.6	10	20.7	140	0.37
c-hME15	0.1167	14.0	3.2	0.5	82.3	15	32.0	228	0.62
c-hME20	0.1167	13.9	3.2	0.5	82.4	20	42.0	280	0.77

All algorithms provide a reasonably good continuous estimates. Variable-structure IMM provides the same continuous estimation quality for half of the runtime cost of the standard IMM but for the price of a significantly degraded mode estimation capability. IMM is slightly better than hybrid estimation (HME and c-hME) in terms of continuous estimation quality. However, hybrid estimation is unmistakably better than IMM in terms of runtime costs. This advantage will be even more dramatic in systems with a larger number of modes. It is evident that the estimation quality of hME and c-hME increases with the number of trajectory hypotheses that are considered during estimation (fringe size). Furthermore, the poor mode estimation of hME1 and c-hME1 suggests that deducing the best estimate only does not provide the potential of our hybrid estimation method. Hybrid estimation with clustering, when compared to its non-clustering variant, performs better over all criteria – continuous estimation, mode estimation, and runtime costs.

The previous experiments used the most likely (MAP) estimate $\hat{\mathbf{x}}_k^{(1)} = \langle \hat{\mathbf{x}}_{d,k}^{(1)}, p_{c,k}^{(1)} \rangle$ of the fringe $\mathcal{X}_k = \{\hat{\mathbf{x}}_k^{(1)}, \ldots, \hat{\mathbf{x}}_k^{(\eta)}\}$ as approximation for the overall hybrid state estimate. This represents the simplest (and fastest) interpretation of the estimation result. In contrast, we can utilize all estimates $\mathcal{X}_k = \{\hat{\mathbf{x}}_k^{(1)}, \ldots, \hat{\mathbf{x}}_k^{(\eta)}\}$ and evaluate the most likely mode (MLM) $\mathbf{m}^{(1)}$, given all η estimates of the fringe. This leads to a hybrid estimate for the hybrid state $\hat{\mathbf{x}}_k \approx \langle \mathbf{m}^{(1)}, p_{c,k} \rangle$, where the continuous estimate merges the fringe estimates at the leading modes (4.79)

$$p_{c,k} = \frac{1}{b_k(\mathbf{m}^{(1)})} \sum_{\nu \mid \hat{\mathbf{x}}_{d,k}^{(\nu)} = \mathbf{m}^{(1)}} b_k^{(\nu)} p_{c,k}^{(\nu)} . \tag{5.3}$$

Table 5.2 records the analysis of the simulation results for our 1-step hybrid estimation algorithm (with clustering). MAP labels the experiments, where the most likely estimate is taken as the hybrid state approximation. MLM denotes the experiments that determine the *most likely mode*, given all η fringe estimates, as the mode estimate and that merge the associated continuous estimates. With FFE we denote the *full fringe estimate* that merges the continuous estimates of all fringe estimates to provide the overall continuous estimate

$$\hat{\mathbf{x}}_{c,k} = \sum_{\nu=1}^{\eta_k} b_k^{(\nu)} \hat{\mathbf{x}}_{c,k}^{(\nu)} . \tag{5.4}$$

There is no significant difference between the possible output strategies, both in terms of the continuous estimate and the mode estimate. Merging trajectories according to the most likely mode or the overall fringe slightly increases the runtime (approximately in the range of 1% to 2% of the overall runtime).

In a third comparison we now rate variants of our N-step hybrid estimation scheme against each other. We vary the window size of N-step hybrid

Table 5.2. Focused hybrid estimation algorithm comparison - output estimates for c-hME.

output estimate	fringe size	rel. error e	mode errors [%] single	double	triple	correct mode [%]
MAP	5	0.116886	17.10	4.40	0.24	78.26
MLM	5	0.116980	16.92	4.50	0.28	78.30
FFE	5	0.116547	-	-	-	-
MAP	10	0.116684	16.70	4.54	0.24	78.52
MLM	10	0.116541	16.50	4.60	0.36	78.54
FFE	10	0.116399	-	-	-	-
MAP	20	0.116699	13.94	3.22	0.46	82.38
MLM	20	0.116778	13.92	3.38	0.42	82.28
FFE	20	0.116399	-	-	-	-

Table 5.3. Focused hybrid estimation algorithm comparison.

window size N	rel. error e	mode errors [%] single	double	triple	correct mode [%]	fringe size	tested hypotheses average	max	rel. runtime
1	0.116684	16.70	4.54	0.24	78.52	10	20.7	140	1.00
2	0.117270	19.12	6.78	0.60	73.50	10	54.5	610	1.90
5	0.117487	27.86	10.88	0.72	60.54	10	73.9	3256	3.18
10	0.117410	31.44	15.42	0.50	52.64	10	109.1	5364	5.21
20	0.119524	39.26	22.44	0.88	37.42	10	140.7	19052	9.21

estimation (again, with clustering). Table 5.3 summarizes the results for window sizes $N = 1, \ldots, 20$, an initial fringe size (at $k - N$) of $\kappa = 5$, and the fixed fringe size (at k) of $\eta = 10$.

One would expect that an increased window size improves the estimation result. However, the experiment shows a contrary result. An increased window size has marginal effects on the continuous estimation. In terms of the mode estimation, however, we can observe a clear trend, where an increased window size leads to a decreased mode prediction capability. An increased window-size puts more and more emphasis on past observations and, as a result, delays the detection of mode changes. This leads to the increased number of wrongly classified modes as we increase the window size of N-step hybrid estimation.

We can summarize the outcome of the experiments above as follows:

- We observe similar prediction quality of our proposed hybrid estimation scheme and standard multiple-model estimation algorithms.
- Our approach is computationally faster and has the prospect to scale better to systems of higher complexity (as we will see in the next example).

- We obtain better estimation quality as we increase the fringe size η, but small fringe sizes ($\eta > 1$) already provide reasonable quality so that we can operate even under stringent computational resource constraints.
- Taking all estimates of the fringe into account only leads to a marginal improvement in estimation quality compared to simply taking the leading estimate.
- An increased window size degrades the mode estimation quality.

Overall, we can conclude that 1-step hybrid estimation with MAP output, which represents the simplest algorithm out of our proposed class of focused hybrid estimation algorithms, provides a reasonably good approximation of the hybrid estimate.

5.2 Advanced Life Support System - BIO-Plex

As an example from the process automation domain, we want to demonstrate our proposed hybrid estimation capability with the BIO-Plex advanced life support system example that was introduced in Section 2. For the scope of this case study, we again restrict our evaluation to the sub-system that deals with CO_2 and O_2 control in one plant growth chamber (PGC), as shown in Fig. 5.3. This sub-system represents a moderately complex automation system that maintains an O_2 concentration at 21 vol.% (± 1 vol.%) and that keeps the CO_2 level at a plant-growth optimal concentration of 1200 ppm during the day phase of the system (20 hours/day). This CO_2 level is unsuitable for humans, hence the gas concentration is lowered to 500 ppm whenever crew members request to enter the chamber for harvesting, re-planting, or other service activities.

Hybrid estimation schemes are key to tracking system operational modes, as well as detecting subtle failures and performing diagnoses. For example,

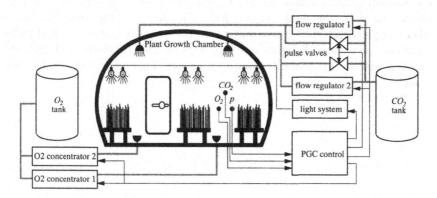

Fig. 5.3. BIO-Plex plant growth chamber subsystem.

we simulate the operational sequence, where the crew requests an entry into the chamber. The system lowers the CO_2 level and unlocks the chamber once the gas concentration reaches a level of approximately 500 ppm. Safety regulations require that the control system inhibits high-volume CO_2 gas injection as long as humans are in the PGC. There are dedicated sensors at the door of the PGC that record entry and exit of crew members for this purpose. However, sensors are known to fail and it is the task of the hybrid monitoring/estimation system to compensate for such a fault and detect the presence of crew members from the slight disturbance of the CO_2 gas concentration that results from the exhaled CO_2. For this purpose it is important that the hybrid monitoring/estimation system not only filters continuous measurements and estimates other physical entities of interest, but also discriminates among the operational and the fault modes of the system. This involves the ability to also identify subtle faults of the system that were not anticipated during the design phase.

In the following we describe the outcome of simulated experiments for operational and fault scenarios. The simulated data is gathered from the execution of a refined and extended subset of NASA JSC's CONFIG model for the BIO-Plex system [74, 47]. Hybrid estimation utilizes a cPHA model that describes the system in terms of 8 components: two redundant flow regulators (FR1, FR2) that provide continuous CO_2 supply, two redundant pulse injection valves (PIV1, PIV2) that provide a means for increasing the CO_2 concentration rapidly, a light system (LS), the plant growth chamber (PGC), and two redundant O_2 concentrators (OC1, OC2) that remove oxygen from the chamber's atmosphere. Figure 5.4 illustrates the interconnection scheme of the PHA components within the cPHA model for the BIO-Plex plant growth chamber system.

The cPHA model captures the behavior of the plant growth chamber in terms of approximately 450,000 modes. Each mode describes the dynamic evolution of the chamber system by a fifth order system of difference equations. For example, the nominal operational condition for plant growth could be characterized by the mode $\mathbf{x}_{d,k} = [m_{r2}, m_{r2}, m_{v1}, m_{v1}, m_{l2}, m_{p2}, m_{o1}, m_{o1}]$, where m_{r2} characterizes a partially open flow regulator, m_{v1} a closed pulse injection valve, m_{l2} all lights on, m_{p2} a plant growth mode at 1200 ppm, and m_{o1} inactive O_2 concentrators, respectively. This mode specifies the raw model:

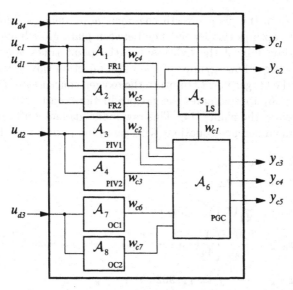

Fig. 5.4. BIO-Plex cPHA model.

$$F_1(m_{r2}) = \{x_{c1,k+1} = 0.5\ u_{c1,k} + v_{c1,k},\ w_{c4} = x_{c1},\ y_{c1} = w_{c4} + v_{c6}\}$$
$$F_2(m_{r2}) = \{x_{c2,k+1} = 0.5\ u_{c1,k} + v_{c2,k},\ w_{c5} = x_{c2},\ y_{c2} = w_{c5} + v_{c7}\}$$
$$F_3(m_{v1}) = \{w_{c2} = 0.0\}$$
$$F_4(m_{v1}) = \{w_{c3} = 0.0\}$$
$$F_5(m_{l2}) = \{w_{c1} = 9028.5\}$$
$$F_6(m_{p2}) = \{x_{c3,k} = x_{c3,k-1} + w_{c8,k-1} + w_{c9,k-1} + v_{c3,k},$$

$$x_{c4,k} = x_{c4,k-1} + \frac{10^6}{x_{c3,k-1}}\ [w_{c8,k-1} - f_1(x_{c4,k-1}, w_{c1,k-1})] + v_{c4,k},$$

$$x_{c5,k} = x_{c5,k-1} + \frac{100}{x_{c3,k-1}}\ [w_{c9,k-1} + f_1(x_{c4,k-1}, w_{c1,k-1})] + v_{c5,k},$$

$$w_{c8} = \frac{1}{44}(w_{c2} + w_{c3} + w_{c4} + w_{c5}),\ w_{c9} = \frac{1}{32}(w_{c6} + w_{c7}),$$

$$y_{c3} = x_{c4} + v_{c8,k},\ y_{c4} = x_{c5} + v_{c9,k},\ y_{c5} = 18.178\ x_{c3} + v_{c10,k}\}$$
$$F_7(m_{o1}) = \{w_{c6} = 0.0\}$$
$$F_8(m_{o1}) = \{w_{c7} = 0.0\}\ ,$$

$$\tag{5.5}$$

where f_1 denotes

$$f_1(x_{c4}, w_{c1}) := 2.3230\ 10^{-7}\ w_{c1}\left[72.0 - 78.89\ e^{-x_{c4}/400.0}\right]. \tag{5.6}$$

$x_{c1,k}$ and $x_{c2,k}$ denote the gas flow (g/min) of flow regulator 1 and 2, respectively. $x_{c3,k}$ captures the total number of gram-moles of gas in the chamber. $x_{c4,k}$ denotes the CO_2 concentration (ppm), and $x_{c5,k}$ denotes the O_2 concentration (vol.%) in the plant growth chamber. $w_{c1,k}$ denotes a multiplicative constant that captures the dependency between plant growth and the photo-synthetic photon flux of the lights above the plant trays. $w_{c2,k}$ and $w_{c3,k}$ denote the CO_2 gas flow (g/min) of the pulse injection valves, and $w_{c6,k}$ and $w_{c7,k}$ denote the O_2 gas flow (g/min) to the oxygen concentrators. The nonlinear equation (5.6) approximates the gas production/consumption rate due to photo-synthesis in the plants [74]. This raw model defines a fifth order system of discrete-time difference equations with sampling-period $T_s = 1\,\text{min}$:

$$x_{c1,k+1} = 0.5\, u_{c1,k} + v_{c1,k}$$
$$x_{c2,k+1} = 0.5\, u_{c1,k} + v_{c2,k}$$
$$x_{c3,k+1} = x_{c3,k} + \frac{1}{44}(x_{c1,k} + x_{c2,k}) + v_{c3,k}$$
$$x_{c4,k+1} = x_{c4,k} + \frac{10^6}{x_{c3,k}}\left[\frac{1}{44}(x_{c1,k} + x_{c2,k}) - f_1(x_{c4,k}, 9028.5)\right] + v_{c4,k}$$
$$x_{c5,k+1} = x_{c5,k} + \frac{100}{x_{c3,k}}f_1(x_{c4,k}, 9028.5) + v_{c5,k} \tag{5.7}$$
$$y_{c1,k} = x_{c1,k} + v_{c6,k}$$
$$y_{c2,k} = x_{c2,k} + v_{c7,k}$$
$$y_{c3,k} = x_{c4,k} + v_{c8,k}$$
$$y_{c4,k} = x_{c5,k} + v_{c9,k}$$
$$y_{c5,k} = 18.178\, x_{c3,k} + v_{c10,k}\,.$$

The first simulated experiment demonstrates the mode tracking ability of hybrid estimation. We simulate the operational mode sequence that the automation system executes after crew members request an entry into the PGC, as shown in Fig. 5.5. The crew requests the entry at time-step $k = 600$. Immediately after this request, the chamber control system lowers the CO_2 concentration to the new set-point of 500 ppm and unlocks the door. The crew opens the door at $k = 745$ and enters the chamber at $k = 750$. They remain inside for 1.5 hours and leave the chamber at $k = 840$. Their entry causes a slight disturbance of the CO_2 concentration at $k = 750$ (see Fig. 5.5a) and the chamber control system reduces the CO_2 gas injection to compensate this additional source of CO_2. The chamber control system also deals with O_2 control concurrently. It maintains an oxygen concentration of 21 ±1 vol.%. Therefore, it activates the oxygen concentrators to withdraw O_2, once the gas level reaches 22 vol.% at $k = 713$. This causes the oxygen concentration to drop until it reaches the lower level of 20 vol.%. This operation also has direct implications onto the pressure in the chamber, as it can be seen in Fig. 5.5b.

We applied 1-step hybrid estimation with a fixed fringe size of 10 estimates and used the estimate $\hat{x}_k^{(1)}$ of the leading hypothesis as the output of the hybrid estimator (MAP). Figure 5.6 shows the mode estimate for the plant growth chamber, as it moves through the operational sequence. We start off with a

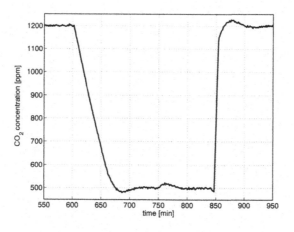

(a) CO_2 concentration

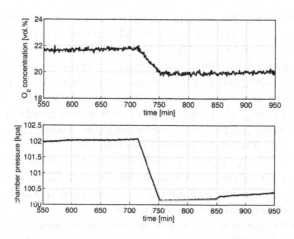

(b) O_2 concentration and pressure

Fig. 5.5. Operational sequence in terms of measurements for the CO_2 and O_2 gas concentration and chamber pressure.

plant growth chamber (PGC) at the plant-growth mode (m_{p2}) and hybrid estimation tracks the transition to the service mode (m_{p6}) via the transition mode m_{p3}, CO_2 control at 500 ppm in an unlocked chamber with a closed door (m_{p4}), and the mode m_{p5} that denotes a situation with an open door but without crew members in the chamber. The presence of crew members in the chamber is captured in terms of the mode m_{p6}. Hybrid estimation prefers this hypothesis from $k = 753$ onward – three minutes after the crew

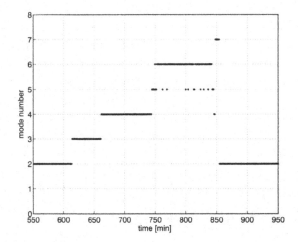

Fig. 5.6. Mode estimation of the plant growth chamber.

entered the chamber. We perform estimation based on the control inputs to the actuators, the observations of the CO_2 and O_2 gas concentration, the pressure, and a discrete door signal that records whether the door is open or closed. Therefore, the detection of the crew entry is done entirely based on the noisy measurements of the gas concentrations and the pressure. This also explains the non-optimal estimate that leads to few miss-classifications between m_{p5} and m_{p6}. Nevertheless, we identify a clear trend that can be used as a backup mechanism for a dedicated sensor that records the entry and exit of crew members. The mode sequence continues with m_{p5} at $k = 743$ (again, 3 minutes late), and the transition mode m_{p7} that moves the chamber back to the plant growth mode m_{p2}. Estimation details, both for the continuous CO_2 concentration and the mode of the chamber, are given in Fig. 5.7. Figure 5.8 shows the mode estimates for all components.

In a second experiment, we repeat the operational sequence but also inject a light fault at $k = 800$. The plant growth chamber has a plant growing area that is arranged in ten shelves that are stacked in three columns – a large center stack with an growing area of 56.7 m^2, and two small side stacks with a growing area of 11.5 m^2 each. We simulate a situation, where the illumination of one side stack fails. The light fault impairs the illumination for about 14% of the plant growing area and reduces the photo synthesis activity. This causes a slight reduction of the CO_2 consumption and the O_2 production of the plants. We observe this event in terms of a bump in the CO_2 concentration, starting at $k = 800$ (Fig. 5.9a). The crew repairs the fault at $k = 936$ and exits the chamber 4 minutes later at $k = 840$. The light fault in one side stack only leads to a slightly modified dynamic behavior since it only harms a small fraction (14%) of the chamber's plant growth area. As a consequence, it is difficult to discriminate between the plant growth mode at full illumination

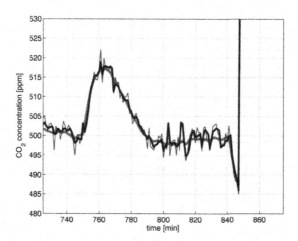

(a) measurement (thin gray), estimate (black) and correct value (thick gray) of the CO_2 concentration

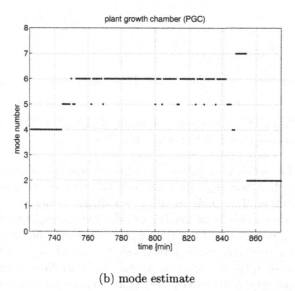

(b) mode estimate

Fig. 5.7. Hybrid estimation detail (experiment 1).

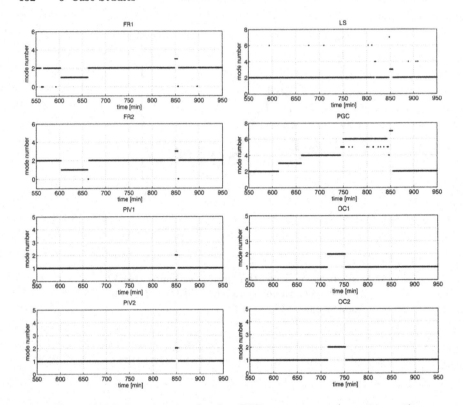

Fig. 5.8. Mode estimation for the cPHA components (experiment 1).

and the predefined fault mode with reduced illumination. This leads to some mis-classifications, as it can be seen in Fig. 5.9b.

The third experiment simulates an unknown operational condition. One of the two redundant flow regulators becomes off-line and drifts slowly towards its closed position. This fault situation is difficult to capture by an explicit fault model as we do not know in advance, whether the regulator drifts towards the fully open or the closed position, nor do we know the magnitude of the drift. A fault of this type, which develops slowly and whose symptom is hidden among the noise in the system, is a typical candidate for our unknown-mode detection capability. Additionally, this fault occurs while there is also a light fault present in the system. This allows us to demonstrate the multiple-fault detection capability of our approach.

Figure 5.10 shows the causal graph of the raw model (5.5). For clarity, we use a slightly simplified graph that omits the noise variables, since they do not change the decomposition. The decomposition of the graph groups the components into a cluster with the first flow regulator ($\{\mathcal{A}_1\}$), one with the second flow regulator ($\{\mathcal{A}_2\}$), and one cluster for the remaining components,

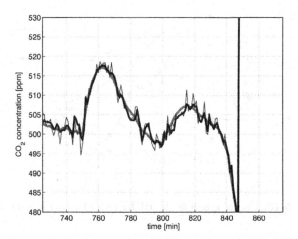

(a) measurement (thin gray), estimate (black) and correct value (thick gray) of the CO_2 concentration

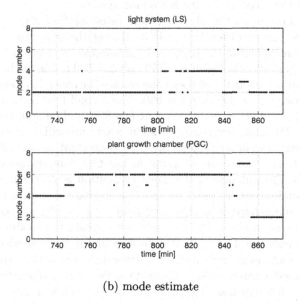

(b) mode estimate

Fig. 5.9. Hybrid estimation detail for operational sequence with light fault (experiment 2).

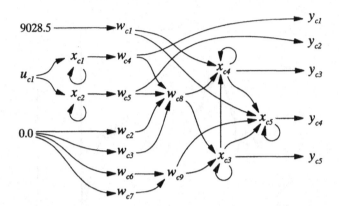

Fig. 5.10. Causal graph of the BIO-Plex cPHA raw model (5.5).

that is, the pulse injection valves, the light system, the chamber, and the oxygen concentrators ($\{A_3, \ldots, A_8\}$). This enables us to estimate the mode and continuous state of the flow regulators independent of the remaining system. For example, an unknown mode in a flow regulator does not impose any constrain on the estimate for the remaining system.

Figure 5.11a shows the observed flow rates for flow regulator 1 and 2 and the CO_2 concentration in the PGC during the experiment. Both flow regulators provide half of the requested gas injection rate up to $k = 850$. The reduction of gas inflow between $k = 750$ and $k = 800$ is due to the crew entry. The light fault occurs at $k = 800$ and one can see the control systems gas inflow adaption between $k = 800$ and $k = 850$. The second flow regulator starts to slowly drift toward its closed position at $k = 850$, and cuts off the gas inflow at $k = 913$. The chamber control system reacts immediately and increases the control signal in order to keep the CO_2 concentration at 1200 ppm. This transient behavior causes an adaption in the CO_2 concentration, as shown in Fig. 5.11b. The outcome of the mode estimation is shown in Fig. 5.12 for the flow regulator and the light system. Our hybrid mode estimation system detects this unmodeled flow regulator fault at $k = 862$ and declares flow regulator 2 to be in the *unknown mode* (we indicate the unknown mode by the mode number 0 in Fig. 5.12). The flow regulator mode *stuck-closed* (m_{r4}) becomes more and more likely as the regulator drifts toward its closed position. Hybrid mode estimation prefers this mode as symptom explanation from $k = 909$ onward, although flow regulator cuts off the inflow a little bit later at $k = 913$.

The light fault at $k = 800$ is detected quickly at $k = 808$ (m_{l4}). Again, the similar dynamics of the plant growth mode with and without the light fault, leads to some mis-classifications, but we can observe a clear trend indicating that there is a light problem in one side stack from $k = 808$ onward.

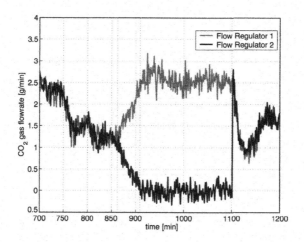

(a) Measured CO_2 inflow rates

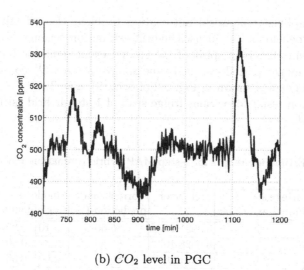

(b) CO_2 level in PGC

Fig. 5.11. Measured CO_2 inflow rates and gas concentration in PGC.

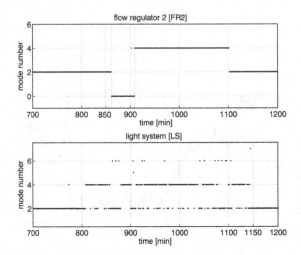

Fig. 5.12. Mode estimates (experiment 3).

This simplified real world automation problem also effectively demonstrates the benefit of our focused hybrid estimation scheme. We performed
an experiment that is in spirit of the second event sequence but tracks the
PGC for a longer period of 1,000 time samples (the CO_2 concentration of
this experiment was shown in Fig. 2.2 of Sect. 2). The Table 5.4 summarizes
the estimation result for various fringe sizes of 1-step hybrid estimation with
MAP output.

Table 5.4. Estimation results for 1-step MAP hybrid estimation and various fringe
sizes.

	fringe size	rel. error	correct mode	mode errors (%)		
type	min / av. / max	e	prediction (%)	single	double	triple
fixed	2	$8.3991\ 10^{-5}$	85.60	13.10	0.80	0.50
fixed	5	$8.0628\ 10^{-5}$	87.60	11.30	0.60	0.50
fixed	10	$8.1291\ 10^{-5}$	86.40	12.60	0.50	0.50
fixed	20	$8.1075\ 10^{-5}$	86.30	12.70	0.50	0.50
auto	2 / 2.92 / 20	$9.3768\ 10^{-5}$	83.30	15.40	0.80	0.50
auto	3 / 5.07 / 100	$8.7115\ 10^{-5}$	83.10	15.60	0.80	0.50

We obtain comparable good results for all fringe sizes. Mode estimation
results suggest that we use a fringe size in the range of 5 to 10 estimates.
We performed also two experiments with automatic fringe size control, based
on the method that was presented in Sect. 4.5.2. For both experiments we
varied the minimal/maximal numbers for the fringe size (2/20, and 3/100, respectively). Compared to the estimation with a fixed fringe size, we observed

Table 5.5. Runtime statistics for 1-step MAP hybrid estimation and various fringe sizes (Pentium P4 1.6GHz, 650MB RAM, Win2000, Allegro Common LISP 6.2).

fringe		tested hypotheses			Filter		average CPU time
type	size	average	max	max/av.	executions	deductions	[seconds] per est. step
fixed	2	48.90	1,244	25.4	139,933	1,111	0.468
fixed	5	120.42	2,340	19.4	345,872	1,126	1.080
fixed	10	243.62	4,908	20.2	700,104	1,167	2.471
fixed	20	487.16	9,911	20.3	1,398,774	1,197	6.441
auto	2.92	189.66	6,784	35.8	532,844	1,167	1.444
auto	5.07	296.01	28,800	97.3	824,983	1,197	1.901

slightly degraded estimation quality, both in terms of the continuous estimate and the mode prediction. Furthermore, automatic fringe size control can lead to a less desirable on-line behavior, where the run-times of the individual estimation steps vary much more, as it is the case for the fixed fringe size experiments. This can be seen in Table 5.5 below, which lists the average and maximal number of estimation hypotheses that are evaluated in the course of the experiment. These numbers directly relate to the computational requirements at each time-step. This evaluation also demonstrates how well our hybrid estimation scheme focuses onto the leading set of trajectories. Take, for example, the estimator that maintains a fixed fringe size of the 10 best estimates. In order to obtain this leading set, it tests in average 243.62, and in the worst case 4,908 hypotheses. Compared to the 451,584 modes of the cPHA, this is less than 0.05% or 1.09% of the possible mode hypotheses for the average and worst case, respectively! The table also records the benefits of the filter decomposition and caching strategy. Estimation with a fringe size of 10 performed 700,104 filter executions in the course of the 1,000 time samples (243.62 × 1,000 = 243,620 filter cluster executions). In order to provide the appropriate filters, our estimator only deduced 1,167 extended Kalman filters. For the majority of cases (98%), it was possible to retrieve the appropriate filter from the cache, thus avoiding many repeated filter deductions. This saves valuable computation time, as well as memory space as we stored at most 0.85% of the 451,584 possible filters for the overall system.

6

Conclusion

6.1 Monograph Revisited

Classical hybrid estimation schemes, such as the family of multiple-model estimation algorithms, do not scale up to the demanding estimation and diagnosis problems that arise in automation systems for modern complex artifacts. This monograph presented a hybrid estimation framework that can remedy this situation. Our proposed framework provides a model-based toolkit for hybrid modeling, on-line model analysis and estimator synthesis, and, of course, hybrid estimation itself.

The core of our framework is a probabilistic, component-based, hybrid modeling paradigm that captures a complex artifact mathematically. The concurrent probabilistic hybrid automaton model (cPHA), as we call it, compactly encodes the vast amount of possible behaviors, both discrete and continuous, of the physical system under investigation. In terms of hybrid estimation, we take ideas from the field of discrete Model-based Reasoning, a sub-field of Artificial Intelligence, and reformulate estimation as a search problem. This enables us to utilize a combination of best-first and beam search techniques that focus the estimation onto highly probably estimation hypotheses, without considering a prohibitively large number of unlikely hypotheses. This reformulation does not only tame the complex hybrid estimation problem, it also leads to an estimation algorithm that is particularly well suited for on-line execution. The search operation provides the estimation hypotheses consecutively, starting with the most likely one. This corresponds to an any-time/any-space formulation and allows us to terminate the computation of an estimate whenever we run short of computation time or memory space. The search-based estimation technique is not the only model-based reasoning tool that we adopt for hybrid systems. Discrete model-based diagnosis provides means for dealing with unmodeled situations, something traditional continuous and hybrid estimation techniques cannot do. This monograph provides a novel approach to incorporate the concept of unknown mode into our hybrid estimation scheme. This approach is based on an on-line model decomposition method that works around the unknown modes of individual components, by constructing a set of concurrent estimators for subsets of the system in which component modes have specified behaviors. The system analysis and filter synthesis are done on-line during the course of hybrid estimation and utilizes efficient algorithms for causal analysis, decomposition, and automated filter deduction. This formulation, together with the search based estimation algorithm provides the means for an overall hybrid estimation engine that is not only capable of dealing with complex systems on-line, it also automates many tasks that are normally done off-line by an experienced control engineer (for example, the Kalman filter design). This is in spirit of a model-based programming paradigm. The design of a hybrid estimator for a particular physical system under investigation only requires one to specify the cPHA model of the system. Plugging this model into our hybrid estimation engine provides the capable estimator sought for. This provides a high level of flexibility, since a modification of the underlying physical artifact only requires an update of the hybrid model or model fragment. The estimation engine automatically deals with the necessary modifications of the underlying filters and captures the system-wide effects of the modification. This flexibility of the estimator, concerning changes in the system's specification and/or topology, opens new perspectives for advanced automation of complex systems. For example, it enables a supervisory control system to dynamically reconfigure the artifact under control in order to adapt to changing environmental conditions and faults. It can do so without having to consider the estimation/monitoring sub-system since the estimator adapts automatically to a changed configuration as well.

6.2 Future Work

The hybrid estimation paradigm, as described in this monograph, deals with some, but surely not all possible issues that arise in the context of hybrid estimation. Possible topics for further research could be:

Modeling: Our modeling paradigm currently limits component models in the form of discrete time difference equations that operate all at the same sampling period T_s. A desirable extension would be to allow systems to be composed of component models that operate with various sampling rates, or even continuously.

cPHA specification and verification: The PHA/cPHA specification as it is used in this work does not constrain the set of equations so that we can ensure that they are 'well formed' and lead to a state space model of the form (3.18). It would be interesting to formulate additional constraints onto the automaton definitions, the compatibility specification, and the automata composition operation so that we can verify: (1) the composition of PHAs into a cPHA does not cause any conflicts regarding the causality in the system, and (2) we obtain a state-space model of the desired form.

Focused search: Currently, we use a combination of standard A∗ and beam search to solve the estimation task. One direction for future research is to utilize the continuous variant of *conflicts*, in order to further focus the underlying search operation. A conflict is a (partial) mode assignment that makes a hypothesis very unlikely. This requires a more general treatment of unknown modes compared to the filter decomposition task introduced above. The decompositional model-based learning system Moriarty [103] introduced continuous variants of conflicts, called *dissents*. We are currently reformulating these dissents for hybrid systems and investigating their incorporation into the underlying search operation. This will lead to an overall framework that unifies our previous work on Livingstone and Titan, Moriarty and hybrid estimation.

Filtering: We apply extended Kalman filtering as the underlying continuous filtering technique. This requires us to restrict the disturbances that act upon the system to additive, white Gaussian noise. This is overly restrictive for many practical applications. However, our hybrid estimation algorithm per se is independent of the particular noise model, it would be interesting to extend our framework to other filtering mechanisms, such as Particle filtering or unscented Kalman filtering. This involves the extension of the automated filter design algorithms to deal with a new class of filters, as well as adapting the probabilistic observation and transition functions accordingly.

Compilation: Currently, our experimental implementation of the hybrid estimation framework performs the systems analysis, decomposition, and filter deduction on-line. Some tasks however, could be done equally well at the compilation phase. For example, one could pre-specify the inputs and outputs of a PHA, whenever the set of equations and the interconnection to the outside world (input specification of the cPHA) enforces unique causal relationships among the variables. Linked to this causality pre-compilation is the possibility to decompose the overall hybrid estimation task not only at the filtering level, but also on the main estimation level, e.g. separating the overall hybrid estimation task into several concurrent estimation tasks of smaller complexity. Shifting some computational tasks into the system's compilation phase can contribute to a faster on-line operation. Nevertheless, it can limit the flexibility in terms of allowing a dynamic reconfiguration of the system and the capability to reason about system-wide interactions. As a consequence, one ought to carefully balance off-line compilation and on-line deduction.

Implementation: We use an experimental implementation of the hybrid estimation engine that is written in Common LISP. It would be worthwhile to re-implement the system within an industrial real-time environment, so that it can operate in conjunction with a process monitoring and supervisory control system.

Autonomous Automation: As indicated above, we see this work as one step towards a novel automation paradigm that robustly controls a complex artifact within a changing or even ill-defined operational environment. This involves dealing with un-anticipated situations and an overwhelmingly large number of possible control strategies. We expect that some of the tools that were developed for hybrid estimation will provide valuable starting points for the dual control problem. On the long run, this would lead to an overall automation system that can dynamically reconfigure itself in order to provide the artifacts functional goal.

References

1. Åström, K.: Introduction to stochastic control theory. Academic Press (1970)
2. Ackerson, G., Fu, K.: On state estimation in switching environments. IEEE Transactions on Automatic Control **15**, 10–17 (1970)
3. Aho, A., Hopcroft, J., Ullman, J.: Data Structures and Algorithms. Addison-Wesley (1983)
4. Alur, R., Dang, T., Esposito, J., Fierro, R., Hur, Y., Ivancic, F., Kumar, V., Lee, I., Mishra, P., Pappas, G., Sokolsky, O.: Hierarchical hybrid modeling of embedded systems. In: T. Henzinger, C. Kirsch (eds.) Proceedings of the first International Workshop on Embedded Software (EMSOFT 2001), *Lecture Notes in Computer Science*, vol. 2211, p. 14 31. Springer Verlag (2001)
5. Alur, R., Dill, D.: A theory of timed automata. Theoretical Computer Science **126**, 183–235 (1994)
6. Alur, R., Grosu, R., Lee, I., Sokolsky, S.: Compositional refinement for hierarchical hybrid systems. In: M. Di Benedetto, A. Sangiovanni-Vincentelli (eds.) Hybrid Systems: Computation and Control, HSCC 2001, *Lecture Notes in Computer Science*, vol. 2034, pp. 33–48. Springer Verlag (2001)
7. Alur, R., Henzinger, T.: Reactive modules. Formal Methods in System Design **15**(1), 7–48 (1999)
8. Alur, R., Henzinger, T., Sontag, E. (eds.): Hybrid Systems III, *Lecture Notes in Computer Science*, vol. 1066. Springer Verlag (1996)
9. Alur, R., Pappas, G. (eds.): Hybrid Systems: Computation and Control, HSCC 2004, *Lecture Notes in Computer Science*, vol. 2993. Springer Verlag (2004)
10. Anderson, B., Moore, J.: Optimal Filtering. Information and System Sciences Series. Prentice Hall (1979)
11. Antsaklis, P., Kohn, W., Lemmon, M., Nerode, A., Sastry, S. (eds.): Hybrid Systems V, *Lecture Notes in Computer Science*, vol. 1567. Springer Verlag (1999)
12. Antsaklis, P., Kohn, W., Nerode, A., Sastry, S. (eds.): Hybrid Systems IV, *Lecture Notes in Computer Science*, vol. 1273. Springer Verlag (1997)
13. Bar-Shalom, Y., Li, X., Kirubarajan, T.: Estimation with Applications to Tracking and Navigation. Wiley (2001)
14. Basseville, M.: Detecting changes in signals and systems - a survey. Automatica **24**(3), 309–326 (1988)

15. Beard, R.V.: Failure accommodation in linear system through self reorganization. Ph.D. thesis, Massachusetts Institute of Technology, Cambridge, MA., USA (1971)
16. Bellman, R.: Dynamic Programming. Princeton University Press, Princeton, New Jersey (1957)
17. Benazera, E., Travé-Massuyès, L., Dague, P.: State tracking of uncertain hybrid concurrent systems. In: Proceedings of the 13th International Workshop on Principles of Diagnosis (DX02), pp. 106–114 (2002)
18. Bengio, Y.: Markovian models for sequential data. Neural Computing Surveys 2, 129–162 (1999)
19. Bertsekas, D.: Dynamic Programming and Optimal Control, vol. 1, 2 edn. Athena Scientific (2000)
20. Blom, H., Bar-Shalom, Y.: The interacting multiple model algorithm for systems with markovian switching coefficients. IEEE Transactions on Automatic Control 33, 780–783 (1988)
21. Bobrow, D. (ed.): Qualitative Reasoning about Physical Systems. MIT-Press (1985)
22. Branicky, M.: Studies in hybrid systems: Modeling, analysis, and control. Ph.D. thesis, Department of Electrical Engineering and Computer Science, MIT (1995)
23. Branicky, M., Borkar, V., Mitter, S.: A unified framework for hybrid control. Tech. Rep. LIDS-P-2239, Laboratory for Information and Decision Systems, MIT (1994)
24. Buchberger, B., Winkler, F. (eds.): Gröbner Bases and Applications. Cambridge Univ. Press (1998)
25. Bujorianu, M.: Extended stochastic hybrid systems and their reachability problem. In: R. Alur, G. Pappas (eds.) Hybrid Systems: Computation and Control, HSCC 2004, Lecture Notes in Computer Science, vol. 2993, pp. 234–249. Springer Verlag (2004)
26. Chen, J., Patton, R.: Robust Model-Based Fault Diagnosis for Dynamic Systems. Kluwer (1999)
27. Clark, R.N., Fosth, D.C., Walton, V.M.: Detecting instrument malfunctions in control systems. IEEE Transactions on Aerospace and Electronic Systems 11(4), 465–473 (1975)
28. Davis, R.: Diagnostic reasoning based on structure and behavior. Artificial Intelligence 24, 347–410 (1984)
29. de Kleer, J., Williams, B.C.: Diagnosing multiple faults. Artificial Intelligence 32(1), 97–130 (1987)
30. de Kleer, J., Williams, B.C.: Diagnosis with behavioral modes. In: Proceedings of IJCAI-89, pp. 1324–1330 (1989)
31. Dearden, R., Clancy, D.: Particle filters for real-time fault detection in planetary rovers. In: Proceedings of the 13th International Workshop on Principles of Diagnosis (DX02), pp. 1–6 (2002)
32. Dearden, R., Hutter, F.: The gaussian particle filter for efficient diagnosis of non-linear systems. In: Proceedings of the 14th International Workshop on Principles of Diagnosis (DX03) (2003)
33. Di Benedetto, M., Sangiovanni-Vincentelli, A. (eds.): Hybrid Systems: Computation and Control, HSCC 2001, Lecture Notes in Computer Science, vol. 2034. Springer Verlag (2001)

34. Dijkstra, E.: A note on two problems in connexion with graphs. Nummerische Mathematik **1**, 269–271 (1959)
35. Doucet, A., de Freitas, N., Gordon, N. (eds.): Sequential Monte Carlo Methods in Practice. Springer Verlag (2001)
36. Elliott, R., Aggoun, L., Moore, J.: Hidden Markov models: estimation and control. Springer Verlag (1995)
37. Faltings, B., Struss, P. (eds.): Recent Advances in Qualitative Physics. MIT Press (1992)
38. Frank, P.: Fault diagnosis in dynamic system via state estimation - a survey. In: System Fault Diagnostics, Reliability and Related Knowledge-based Approaches, pp. 35–98. D. Reidel Press (1987)
39. Frank, P.: Enhancement of robustness in observer-based fault detection. International Journal of Control **59**(4), 955–981 (1994)
40. Funiak, S., Williams, B.: Multi-modal particle filtering for hybrid systems with autonomous mode transitions. In: Proceedings of the 14th International Workshop on Principles of Diagnosis (DX03) (2003)
41. Gehin, A., Assas, M., Staroswiecki, M.: Structural analysis of system reconfigurability. In: Preprints of the 4th IFAC SAFEPROCESS Symposium, vol. 1, pp. 292–297 (2000)
42. Gelb, A.: Applied Optimal Estimation, 15th printing, 1999 edn. MIT Press (1974)
43. Gertler, J.: Fault detection and isolation using parity relations. Control Engineering Practice **5**(5), 653–661 (1997)
44. Glover, W., Lygeros, J.: A stochastic hybrid model for air traffic control simulation. In: R. Alur, G. Pappas (eds.) Hybrid Systems: Computation and Control, HSCC 2004, *Lecture Notes in Computer Science*, vol. 2993, pp. 372–386. Springer Verlag (2004)
45. Grewal, M., Andrews, A.: Kalman Filtering: Theory and Practice, 2 edn. Prentice Hall (2001)
46. Hamscher, W., Console, L., de Kleer, J. (eds.): Readings in Model-Based Diagnosis. Morgan Kaufmann (1992)
47. Hanford, A.: Advanced life support baseline values and assumptions document. Tech. Rep. CTSD-ADV-484, NASA, Johnson Space Center, Houston (2002)
48. Hanlon, P., Maybeck, P.: Multiple-model adaptive estimation using a residual correlation kalman filter bank. IEEE Transactions on Aerospace and Electronic Systems **36**(2), 393–406 (2000)
49. Hart, P., Nilsson, N., Raphael, B.: A formal basis for the heuristic determination of minimum cost paths. IEEE Transactions on Systems, Man, and Cybernetics **4**(2), 100–107 (1968)
50. Hart, P., Nilsson, N., Raphael, B.: Correction to "A formal basis for the heuristic determination of minimum cost paths". SIGART Newsletter **37**, 28–29 (1972)
51. Henzinger, T.: Masaccio: A formal model for embedded components. In: J. van Leeuwen, O. Watanabe, M. Hagiya, P. Mosses, T. Ito (eds.) Proceedings of the First IFIP International Conference on Theoretical Computer Science, *Lecture Notes in Computer Science*, vol. 1872, pp. 549–563. Springer Verlag (2000)
52. Henzinger, T., Sastry, S. (eds.): Hybrid Systems: Computation and Control, HSCC 1998, *Lecture Notes in Computer Science*, vol. 1386. Springer-Verlag (1998)

53. Hofbaur, M.W.: Hybrid estimation and its role in automation. Habilitationss-chrift, Faculty of Electrical Engineering, Graz University of Technology, Austria (2003)

54. Hofbaur, M.W., Williams, B.C.: Hybrid diagnosis with unknown behavioral modes. In: Proceedings of the 13th International Workshop on Principles of Diagnosis (DX02), pp. 97–105 (2002)

55. Hofbaur, M.W., Williams, B.C.: Mode estimation of probabilistic hybrid systems. In: C. Tomlin, M. Greenstreet (eds.) Hybrid Systems: Computation and Control, HSCC 2002, *Lecture Notes in Computer Science*, vol. 2289, pp. 253–266. Springer Verlag (2002)

56. Hofbaur, M.W., Williams, B.C.: Hybrid estimation of complex systems. IEEE Transactions on Systems, Man, and Cybernetics - Part B: Cybernetics **34**(5), 2178–2191 (2004)

57. Hu, J., Lygeros, J., Sastry, S.: Towards a theory of stochastic hybrid systems. In: N. Lynch, B. Krogh (eds.) Hybrid Systems: Computation and Control, HSCC 2000, *Lecture Notes in Computer Science*, vol. 1790, pp. 160–173. Springer Verlag (2000)

58. Isermann, R.: Process fault detection based on modeling and estimation methods - a survey. Automatica **20**(4), 384–404 (1984)

59. Jones, H.L.: Failure detection in linear systems. Ph.D. thesis, Massachusetts Institute of Technology, Cambridge, MA., USA (1973)

60. Julier, S., Uhlmann, J.: A new extension of the kalman filter to nonlinear systems. In: Proceedings of the SPIE AeroSense Symposium, Orlando, FL (1997)

61. Kalman, R.: A new approach to linear filtering and prediction problems. ASME Transactions, Journal of Basic Engineering **82**, 35–50 (1960)

62. Kleissl, W.: Structural analysis of hybrid systems. MSc thesis, Institute of Automation and Control, Graz University of Technology, Graz, Austria (2002)

63. Koutsoukos, X., Antsaklis, P., Stiver, J., Lemmon, M.: Supervisory control of hybrid systems. Proceedings of the IEEE **88**(7), 1026–1049 (2000)

64. Koutsoukos, X., Kurien, J., Zhao, F.: Estimation of distributed hybrid systems using particle filtering methods. In: O. Maler, A. Pnueli (eds.) Hybrid Systems: Computation and Control, HSCC 2003, *Lecture Notes in Computer Science*, vol. 2623, pp. 298–313. Springer Verlag (2003)

65. Lerner, U., Parr, R., Koller, D., Biswas, G.: Bayesian fault detection and diagnosis in dynamic systems. In: Proceedings of the 7th National Conference on Artificial Intelligence (AAAI'00) (2000)

66. Li, P., Kadirkamanathan, V.: Particle filtering based likelyhood ratio approach to fault diagnosis in nonlinear stochastic systems. IEEE Transactions on Systems, Man, and Cybernetics - Part C **31**(3), 337–343 (2001)

67. Li, X.: Multiple-model estimation with variable structure - part II: Model-set adaption. IEEE Transactions on Automatic Control **45**, 2047–2060 (2000)

68. Li, X., Bar-Shalom, Y.: Multiple-model estimation with variable structure. IEEE Transactions on Automatic Control **41**, 478–493 (1996)

69. Li, X., Zhi, X., Zhang, Y.: Multiple-model estimation with variable structure - part III: Model-group switching algorithm. IEEE Transactions on Aerospace and Electronic Systems **35**(1), 225–241 (1999)

70. Lunze, J.: On the markov property of quantised state measurement sequences. Automatica **34**(11), 1439–1444 (1998)

71. Lynch, N., Krogh, B. (eds.): Hybrid Systems: Computation and Control, HSCC 2000, *Lecture Notes in Computer Science*, vol. 1790. Springer Verlag (2000)
72. Lynch, N., Segala, R., Vaandrager, F., Weinberg, H.: Hybrid I/O automata. Tech. Rep. CSI-R9907, Computing Science Institute Nijmegen (1999)
73. Maler, O., Pnueli, A. (eds.): Hybrid Systems: Computation and Control, HSCC 2003, *Lecture Notes in Computer Science*, vol. 2623. Springer Verlag (2003)
74. Malin, J.T., Fleming, L., Hatfield, T.R.: Interactive simulation-based testing of product gas transfer integrated monitoring and control software for the lunar mars life support phase III test. In: SAE 28th International Conference on Environmental Systems, Danvers MA (1998)
75. Maybeck, P., Stevens, R.: Reconfigurable flight control via multiple model adaptive control methods. IEEE Transactions on Aerospace and Electronic Systems **27**(3), 470–480 (1991)
76. McGee, L., Schmidt, S.: Discovery of the Kalman filter as a practical tool for aerospace and industry. Tech. Rep. 86847, NASA AMES (1985)
77. McIlraith, S.: Diagnosing hybrid systems: a bayseian model selection approach. In: Proceedings of the 11th International Workshop on Principles of Diagnosis (DX00), pp. 140–146 (2000)
78. McIlraith, S., Biswas, G., Clancy, D., Gupta, V.: Towards diagnosing hybrid systems. In: Proceedings of the 10th International Workshop on Principles of Diagnosis (DX99) (1999)
79. Milne, R.: Strategies for diagnosis. IEEE Transactions on Systems, Man, and Cybernetics **17**(3), 333–339 (1987)
80. Mosterman, P., Biswas, G.: Building hybrid observers for complex dynamic systems using model abstractions. In: F. Vaandrager, J. Schuppen (eds.) Hybrid Systems: Computation and Control, HSCC 1999, *Lecture Notes in Computer Science*, vol. 1569, pp. 178–192. Springer Verlag (1999)
81. Nancy Lynch Roberto Segala, F.V.: Hybrid I/O automata revisited. In: M. Di Benedetto, A. Sangiovanni-Vincentelli (eds.) Hybrid Systems: Computation and Control, HSCC 2001, *Lecture Notes in Computer Science*, vol. 2034, pp. 403–417. Springer Verlag (2001)
82. Narasimhan, S., Biswas, G.: An approach to model-based diagnosis of hybrid systems. In: C. Tomlin, M. Greenstreet (eds.) Hybrid Systems: Computation and Control, HSCC 2002, *Lecture Notes in Computer Science*, vol. 2289, pp. 308–322. Springer Verlag (2002)
83. Nayak, P.: Automated Modelling of Physical Systems. Lecture Notes in Artificial Intelligence. Springer Verlag (1995)
84. Papoulis, A.: Probability, Random Variables, and Stochastic Processes, 2 edn. Mc Graw Hill (1984)
85. Platzer, M., Rinner, B., Weiss, R.: A computer architecture to support qualitative simulation in industrial applications. e&i **114**(1), 13–18 (1997)
86. Rabiner, L.: A tutorial on hidden markov models and selected applications in speech recognition. Proceedings of the IEEE **77**(2), 257–286 (1989)
87. Reinschke, K.: Multivariable Control - A Graph-theoretic Approach, *Lecture Notes in Control and Information Sciences*, vol. 108. Springer Verlag (1988)
88. Rinner, B.: Monitoring and diagnosis of technical systems. Habilitationsschrift, Faculty of Electrical Engineering, Graz University of Technology (2001)
89. Russel, S., Norvig, P.: Artificial Intelligence, 2 edn. Prentice-Hall (2003)

90. Sachenbacher, M., Malik, A., Struss, P.: From electrics to emissions: Experiences in applying model-based diagnosis to real problems in real cars. In: Proceedings of the 9th International Workshop on Principles of Diagnosis (DX98) (1998)
91. Sontag, E.: Mathematical Control Theory: Deterministic Finite Dimensional Systems, 2 edn. Springer Verlag (1998)
92. Stengel, R.: Optimal Control and Estimation. Dover (1994)
93. Tomlin, C., Greenstreet, M. (eds.): Hybrid Systems: Computation and Control, HSCC 2002, *Lecture Notes in Computer Science*, vol. 2289. Springer Verlag (2002)
94. Tong, Y.: Probability Inequalities in Multivariate Distributions. Academic Press (1980)
95. Travé-Massuyès, L., Milne, R.: Gas-turbine condition monitoring using qualitative model-based diagnosis. IEEE Expert **12**(3), 22–31 (1997)
96. Travé-Massuyès, L., Pons, R.: Causal ordering for multiple mode systems. In: Proceedings of the 11th International Workshop on Qualitative Reasoning (QR97), pp. 203–214 (1997)
97. Tugnait, J.: Detection and estimation for abruptly changing systems. Automatica **18**, 607–615 (1982)
98. Vaandrager, F., Schuppen, J. (eds.): Hybrid Systems: Computation and Control, HSCC 1999, *Lecture Notes in Computer Science*, vol. 1569. Springer Verlag (1999)
99. Verma, V., Gordon, G., Simmons, R., Thrun, S.: Real-time fault diagnosis. IEEE Robotics and Automation Magazine **11**(2), 56–66 (2004)
100. Weld, D., de Kleer, J. (eds.): Qualitative Reasoning about Physical Systems. Morgan Kaufmann (1990)
101. Williams, B., Ingham, M., Chung, S., Elliott, P., Hofbaur, M., Sullivan, G.: Model-based programming of fault-aware systems. AI Magazine **24**(4), 61–75 (2003)
102. Williams, B.C., Hofbaur, M.W., Jones, T.: Mode estimation of probabilistic hybrid systems. Tech. Rep. SSL 6-01, MIT Space Systems Laboratory (2001)
103. Williams, B.C., Millar, B.: Decompositional, model-based learning and its analogy to diagnosis. In: Proceedings of the 15th National Conference on Artificial Intelligence (AAAI-98) (1998)
104. Williams, B.C., Nayak, P.: A model-based approach to reactive self-configuring systems. In: Proceedings of the 13th National Conference on Artificial Intelligence (AAAI-96) (1996)
105. Williams, B.C., Nayak, P., Muscettola, N.: Remote agent: To boldly go where no AI system has gone before. Artificial Intelligence **103**(1-2), 5–48 (1998)
106. Williams, B.C., Ragno, R.J.: Conflict-directed A* and its role in model-based embedded systems. Journal of Discrete Applied Math (2003)
107. Willsky, A.S.: A survey of design methods for failure detection in dynamic systems. Automatica **12**(6), 601–611 (1974)
108. Winston, P.: Artificial Intelligence, 3 edn. Addison-Wesley (1993)
109. Zhang, Q.: Hybrid filtering for linear systems with non-gaussian disturbances. IEEE Transactions on Automatic Control **45**, 50–61 (2000)
110. Zhao, F., Koutsoukos, X., Haussecker, H., Reich, J., Cheung, P.: Distributed monitoring of hybrid systems: A model-directed approach. In: Proceedings of the International Joint Conference on Artificial Intelligence (IJCAI'01), pp. 557–564 (2001)

Lecture Notes in Control and Information Sciences

Edited by M. Thoma and M. Morari

Further volumes of this series can be found on our homepage:
springeronline.com

Vol. 318: Eli Gershon; Uri Shaked; Isaac Yaesh
H_∞ Control and Estimation of State-muliplicative
Linear Systems
256 p. 2005 [3-540-1-85233-997-7]

Vol. 317: Chuan Ma; Murray Wonham
Nonblocking Supervisory Control of State Tree
Structures
208 p. 2005 [3-540-25069-7]

Vol. 316: R.V. Patel, F. Shadpey
Control of Redundant Robot Manipulators
224 p. 2005 [3-540-25071-9]

Vol. 315: Herbordt, W.
Sound Capture for Human/Machine Interfaces:
Practical Aspects of Microphone Array Signal Processing
286 p. 2005 [3-540-23954-5]

Vol. 314: Gil', M.I.
Explicit Stability Conditions for Continuous Systems
193 p. 2005 [3-540-23984-7]

Vol. 313: Li, Z.; Soh, Y.; Wen, C.
Switched and Impulsive Systems
277 p. 2005 [3-540-23952-9]

Vol. 312: Henrion, D.; Garulli, A. (Eds.)
Positive Polynomials in Control
313 p. 2005 [3-540-23948-0]

Vol. 311: Lamnabhi-Lagarrigue, F.; Loría, A.;
Panteley, V. (Eds.)
Advanced Topics in Control Systems Theory
294 p. 2005 [1-85233-923-3]

Vol. 310: Janczak, A.
Identification of Nonlinear Systems Using Neural
Networks and Polynomial Models
197 p. 2005 [3-540-23185-4]

Vol. 309: Kumar, V.; Leonard, N.; Morse, A.S. (Eds.)
Cooperative Control
301 p. 2005 [3-540-22861-6]

Vol. 308: Tarbouriech, S.; Abdallah, C.T.; Chiasson, J. (Eds.)
Advances in Communication Control Networks
358 p. 2005 [3-540-22819-5]

Vol. 307: Kwon, S.J.; Chung, W.K.
Perturbation Compensator based Robust Tracking
Control and State Estimation of Mechanical Systems
158 p. 2004 [3-540-22077-1]

Vol. 306: Bien, Z.Z.; Stefanov, D. (Eds.)
Advances in Rehabilitation
472 p. 2004 [3-540-21986-2]

Vol. 305: Nebylov, A.
Ensuring Control Accuracy
256 p. 2004 [3-540-21876-9]

Vol. 304: Margaris, N.I.
Theory of the Non-linear Analog Phase Locked Loop
303 p. 2004 [3-540-21339-2]

Vol. 303: Mahmoud, M.S.
Resilient Control of Uncertain Dynamical Systems
278 p. 2004 [3-540-21351-1]

Vol. 302: Filatov, N.M.; Unbehauen, H.
Adaptive Dual Control: Theory and Applications
237 p. 2004 [3-540-21373-2]

Vol. 301: de Queiroz, M.; Malisoff, M.; Wolenski, P. (Eds.)
Optimal Control, Stabilization and Nonsmooth Analysis
373 p. 2004 [3-540-21330-9]

Vol. 300: Nakamura, M.; Goto, S.; Kyura, N.; Zhang, T.
Mechatronic Servo System Control
Problems in Industries and their Theoretical Solutions
212 p. 2004 [3-540-21096-2]

Vol. 299: Tarn, T.-J.; Chen, S.-B.; Zhou, C. (Eds.)
Robotic Welding, Intelligence and Automation
214 p. 2004 [3-540-20804-6]

Vol. 298: Choi, Y.; Chung, W.K.
PID Trajectory Tracking Control for Mechanical Systems
127 p. 2004 [3-540-20567-5]

Vol. 297: Damm, T.
Rational Matrix Equations in Stochastic Control
219 p. 2004 [3-540-20516-0]

Vol. 296: Matsuo, T.; Hasegawa, Y.
Realization Theory of Discrete-Time Dynamical Systems
235 p. 2003 [3-540-40675-1]

Vol. 295: Kang, W.; Xiao, M.; Borges, C. (Eds)
New Trends in Nonlinear Dynamics and Control,
and their Applications
365 p. 2003 [3-540-10474-0]

Vol. 294: Benvenuti, L.; De Santis, A.; Farina, L. (Eds)
Positive Systems: Theory and Applications (POSTA 2003)
414 p. 2003 [3-540-40342-6]

Vol. 293: Chen, G. and Hill, D.J.
Bifurcation Control
320 p. 2003 [3-540-40341-8]

Vol. 292: Chen, G. and Yu, X.
Chaos Control
380 p. 2003 [3-540-40405-8]

Vol. 291: Xu, J.-X. and Tan, Y.
Linear and Nonlinear Iterative Learning Control
189 p. 2003 [3-540-40173-3]

Vol. 290: Borrelli, F.
Constrained Optimal Control
of Linear and Hybrid Systems
237 p. 2003 [3-540-00257-X]

Vol. 289: Giarré, L. and Bamieh, B.
Multidisciplinary Research in Control
237 p. 2003 [3-540-00917-5]

Vol. 288: Taware, A. and Tao, G.
Control of Sandwich Nonlinear Systems
393 p. 2003 [3-540-44115-8]

Vol. 287: Mahmoud, M.M.; Jiang, J.; Zhang, Y.
Active Fault Tolerant Control Systems
239 p. 2003 [3-540-00318-5]

Vol. 286: Rantzer, A. and Byrnes C.I. (Eds)
Directions in Mathematical Systems
Theory and Optimization
399 p. 2003 [3-540-00065-8]

Vol. 285: Wang, Q.-G.
Decoupling Control
373 p. 2003 [3-540-44128-X]

Vol. 284: Johansson, M.
Piecewise Linear Control Systems
216 p. 2003 [3-540-44124-7]

Vol. 283: Fielding, Ch. et al. (Eds)
Advanced Techniques for Clearance of
Flight Control Laws
480 p. 2003 [3-540-44054-2]

Vol. 282: Schröder, J.
Modelling, State Observation and
Diagnosis of Quantised Systems
368 p. 2003 [3-540-44075-5]

Vol. 281: Zinober A.; Owens D. (Eds)
Nonlinear and Adaptive Control
416 p. 2002 [3-540-43240-X]

Vol. 280: Pasik-Duncan, B. (Ed)
Stochastic Theory and Control
564 p. 2002 [3-540-43777-0]

Vol. 279: Engell, S.; Frehse, G.; Schnieder, E. (Eds)
Modelling, Analysis, and Design of Hybrid Systems
516 p. 2002 [3-540-43812-2]

Vol. 278: Chunling D. and Lihua X. (Eds)
H_∞ Control and Filtering of
Two-dimensional Systems
161 p. 2002 [3-540-43329-5]

Vol. 277: Sasane, A.
Hankel Norm Approximation
for Infinite-Dimensional Systems
150 p. 2002 [3-540-43327-9]

Vol. 276: Bubnicki, Z.
Uncertain Logics, Variables and Systems
142 p. 2002 [3-540-43235-3]

Vol. 275: Ishii, H.; Francis, B.A.
Limited Data Rate in Control Systems with Networks
171 p. 2002 [3-540-43237-X]

Vol. 274: Yu, X.; Xu, J.-X. (Eds)
Variable Structure Systems:
Towards the 21st Century
420 p. 2002 [3-540-42965-4]

Vol. 273: Colonius, F.; Grüne, L. (Eds)
Dynamics, Bifurcations, and Control
312 p. 2002 [3-540-42560-9]

Vol. 272: Yang, T.
Impulsive Control Theory
363 p. 2001 [3-540-42296-X]

Vol. 271: Rus, D.; Singh, S.
Experimental Robotics VII
585 p. 2001 [3-540-42104-1]

Vol. 270: Nicosia, S. et al.
RAMSETE
294 p. 2001 [3-540-42090-8]

Vol. 269: Niculescu, S.-I.
Delay Effects on Stability
400 p. 2001 [1-85233-291-316]

Vol. 268: Moheimani, S.O.R. (Ed)
Perspectives in Robust Control
390 p. 2001 [1-85233-452-5]

Vol. 267: Bacciotti, A.; Rosier, L.
Liapunov Functions and Stability in Control Theory
224 p. 2001 [1-85233-419-3]

Vol. 266: Stramigioli, S.
Modeling and IPC Control of Interactive Mechanical
Systems – A Coordinate-free Approach
296 p. 2001 [1-85233-395-2]

Vol. 265: Ichikawa, A.; Katayama, H.
Linear Time Varying Systems and Sampled-data Systems
376 p. 2001 [1-85233-439-8]

Vol. 264: Baños, A.; Lamnabhi-Lagarrigue, F.;
Montoya, F.J
Advances in the Control of Nonlinear Systems
344 p. 2001 [1-85233-378-2]

Vol. 263: Galkowski, K.
State-space Realization of Linear 2-D Systems with
Extensions to the General nD (n>2) Case
248 p. 2001 [1-85233-410-X]

Vol. 262: Dixon, W.; Dawson, D.M.; Zergeroglu, E.;
Behal, A.
Nonlinear Control of Wheeled Mobile Robots
216 p. 2001 [1-85233-414-2]

Vol. 261: Talebi, H.A.; Patel, R.V.; Khorasani, K.
Control of Flexible-link Manipulators
Using Neural Networks
168 p. 2001 [1-85233-409-6]

Vol. 260: Kugi, A.
Non-linear Control Based on Physical Models
192 p. 2001 [1-85233-329-4]